ALTERNATIVES
FOR HIGH-LEVEL
WASTE SALT
PROCESSING
AT THE
SAVANNAH
RIVER SITE

Committee on Cesium Processing Alternatives for High-Level Waste
at the Savannah River Site

Board on Radioactive Waste Management
Board on Chemical Sciences and Technology

National Research Council

NATIONAL ACADEMY PRESS
Washington, D.C.

Support for this study was provided by the U.S. Department of Energy under Grant No.DE-FC01-99EW59049. All opinions, findings, conclusions, or recommendations expressed herein are those of the authors and do not necessarily reflect the views of the U.S. Department of Energy.

International Standard Book Number 0-309-07194-1

Additional copies of this report are available from:

National Academy Press
2101 Constitution Avenue, N.W.
Box 285
Washington, DC 20055
800-624-6242
202-334-3313 (in the Washington Metropolitan Area)
http://www.nap.edu

THE NATIONAL ACADEMIES

National Academy of Sciences
National Academy of Engineering
Institute of Medicine
National Research Council

The **National Academy of Sciences** is a private, nonprofit, self-perpetuating society of distinguished scholars engaged in scientific and engineering research, dedicated to the furtherance of science and technology and to their use for the general welfare. Upon the authority of the charter granted to it by the Congress in 1863, the Academy has a mandate that requires it to advise the federal government on scientific and technical matters. Dr. Bruce M. Alberts is president of the National Academy of Sciences.

The **National Academy of Engineering** was established in 1964, under the charter of the National Academy of Sciences, as a parallel organization of outstanding engineers. It is autonomous in its administration and in the selection of its members, sharing with the National Academy of Sciences the responsibility for advising the federal government. The National Academy of Engineering also sponsors engineering programs aimed at meeting national needs, encourages education and research, and recognizes the superior achievements of engineers. Dr. William A. Wulf is president of the National Academy of Engineering.

The **Institute of Medicine** was established in 1970 by the National Academy of Sciences to secure the services of eminent members of appropriate professions in the examination of policy matters pertaining to the health of the public. The Institute acts under the responsibility given to the National Academy of Sciences by its congressional charter to be an adviser to the federal government and, upon its own initiative, to identify issues of medical care, research, and education. Dr. Kenneth I. Shine is president of the Institute of Medicine.

The **National Research Council** was organized by the National Academy of Sciences in 1916 to associate the broad community of science and technology with the Academy's purposes of furthering knowledge and advising the federal government. Functioning in accordance with general policies determined by the Academy, the Council has become the principal operating agency of both the National Academy of Sciences and the National Academy of Engineering in providing services to the government, the public, and the scientific and engineering communities. The Council is administered jointly by both Academies and the Institute of Medicine. Dr. Bruce M. Alberts and Dr. William A. Wulf are chairman and vice chairman, respectively, of the National Research Council.

COMMITTEE ON CESIUM PROCESSING ALTERNATIVES FOR HIGH-LEVEL WASTE AT THE SAVANNAH RIVER SITE

MILTON LEVENSON, *Chair*, Bechtel International (retired), Menlo Park, California
GREGORY R. CHOPPIN, *Vice Chair,* Florida State University, Tallahassee
JOHN BERCAW, California Institute of Technology, Pasadena
DARYLE H. BUSCH, University of Kansas, Lawrence
TERESA FRYBERGER, Brookhaven National Laboratory, Upton, New York
GEORGE E. KELLER II, Union Carbide Corporation (retired), South Charleston, West Virginia
MATTHEW KOZAK, Monitor Scientific, LLC, Denver, Colorado
ALFRED P. SATTELBERGER, Los Alamos National Laboratory, Los Alamos, New Mexico
BARRY E. SCHEETZ, The Pennsylvania State University, University Park
MARTIN J. STEINDLER, Argonne National Laboratory (retired), Downers Grove, Illinois

Staff

ROBERT S. ANDREWS, Senior Staff Officer, Board on Radioactive Waste Management
KEVIN D. CROWLEY, Director, Board on Radioactive Waste Management
DOUGLAS J. RABER, Director, Board on Chemical Sciences and Technology
JOHN R. WILEY, Senior Staff Officer, Board on Radioactive Waste Management
LATRICIA C. BAILEY, Senior Project Assistant
MATTHEW BAXTER-PARROTT, Project Assistant
SUZANNE STACKHOUSE, Project Assistant

BOARD ON RADIOACTIVE WASTE MANAGEMENT

Acknowledgements

This fast-track study could not have been completed without the assistance of many individuals and organizations. The committee especially wishes to acknowledge and thank Steve Piccolo (Westinghouse Savannah River Company [WSRC]), Roy Schepens (U.S. Department of Energy [DOE]), Kurt Fisher (DOE), and Kenneth Lang (DOE), who served as liaisons to the committee from Savannah River and DOE Headquarters and provided information and briefings during the course of this study; Robert Jones (WSRC), who served as the committee's main point of contact to the salt processing project at Savannah River and kept track of the committee's sometimes insatiable requests for information; and the other individuals listed in Appendix C, who provided briefings to the committee, in some cases on very short notice, during its two open meetings in Augusta, Georgia.

The committee also wishes to thank the following staff of the National Research Council's Board on Radioactive Waste Management and Board on Chemical Sciences and Technology for a strong team effort to support this project: Robert Andrews, Kevin Crowley, Doug Raber, and John Wiley for organizing the project and assisting with the preparation of the committee's reports; and Latricia Bailey, Matthew Baxter-Parrott, and Suzanne Stackhouse for assistance with research and logistics.

List of Report Reviewers

This report has been reviewed in draft form by individuals chosen for their diverse perspectives and technical expertise, in accordance with procedures approved by the National Research Council (NRC) Report Review Committee. The purpose of this independent review is to provide candid and critical comments that will assist the institution in making the published report as sound as possible and to ensure that the report meets institutional standards for objectivity, evidence, and responsiveness to the study charge. The content of the review comments and draft manuscript remain confidential to protect the integrity of the deliberative process. We wish to thank the following individuals for their participation in the review of this report:

Robert M. Bernero, U.S. Nuclear Regulatory Commission (retired)
Mary L. Good, Venture Capitol Investors
Edward Lahoda, Westinghouse Science and Technology Department
Dade W. Moeller, Dade Moeller & Associates, Inc.
Kenneth N. Raymond, University of California, Berkeley
Vincent Van Brunt, University of South Carolina
Stephen Yates, University of Kentucky
Edwin L. Zebroski, Elgis Consulting

Although the reviewers listed above have provided many constructive comments and suggestions, they were not asked to endorse the conclusions or recommendations, nor did they see the final draft of the report before its release. The review of this report was overseen by E-an Zen, appointed by the Commission on Geosciences, Environment, and Resources, and Royce W. Murray, appointed by the Report Review Committee, who were responsible for making certain that an independent examination of this report was carried out in accordance with NRC procedures and that all review comments were carefully considered. Responsibility for the final content of this report rests entirely with the authoring committee and the NRC.

Contents

Summary

At the request of the U.S. Department of Energy (DOE), the National Research Council formed a committee to provide an independent technical review of alternatives selected by representatives at the Savannah River Site (SRS), South Carolina, for processing the high-level radioactive waste (HLW) salt solutions stored at the site. This review, which is summarized in this report, addresses the following four charges in the committee's statement of task:

- Was an appropriately comprehensive set of cesium partitioning alternatives identified and are there other alternatives that should be explored?
- Was the process used to screen the alternatives technically sound and did its application result in the selection of appropriate preferred alternatives?
- Are there significant barriers to the implementation of any of the preferred alternatives, taking into account their state of development and their ability to be integrated into the existing SRS HLW system?
- Are the planned research and development (R&D) activities, including pilot-scale testing, adequate to support implementation of a single preferred alternative?

HLW at Savannah River is currently being stored in 48 below-grade tanks in the F and H Areas at the site. Currently, the solid (sludge) portion of the tank wastes is being removed and immobilized in borosilicate glass in the Defense Waste Processing Facility (DWPF) for eventual disposal in a geological repository. What remains in the tanks after the removal of sludge is a highly alkaline salt that is present in both liquid (or supernate) and solid (or saltcake) forms. The salt contains cesium, strontium, actinides such as plutonium and neptunium, and other radionuclides. This HLW salt will be put into solution and then processed to remove these radionuclides for immobilization in glass in the DWPF. The remaining "decontaminated" salt solutions will be immobilized in grout at the site. The process originally developed to accomplish this HLW processing—in-tank precipitation (ITP) using monosodium titanate to remove strontium and actinides and sodium tetraphenylborate (TPB) to remove cesium—encountered unexpected problems when operations first started on a large scale with real waste in 1995. Savannah River abandoned the ITP process in 1998 and is now attempting to select an alternative processing option.

After an extensive review of possible processing options, Savannah River selected a primary alternative (a variation of the ITP process known as *small tank TPB precipitation*) and a backup option (an ion exchange process using crystalline silicotitanate, or *CST ion exchange*). At this point DOE requested that the National Research Council undertake the current study to review both the selection process and the processing options themselves, which included the primary and backup options along with two other options that were eliminated by SRS in the final stages of screening: *caustic side solvent extraction* and *direct grout*. Information on these processes is presented in Chapters 3-7 of this report.

The committee issued an interim report on October 14, 1999 (National Research Council, 1999b; see Appendix B), that provided a partial response to the charges in the committee's task statement. This final report provides a more comprehensive response to all four of the charges in the statement of task. Explicit responses to each of those charges are provided below. More details on these recommendations can be found in the main body of the report.

Charge 1: Was an appropriately comprehensive set of cesium partitioning alternatives identified and are there other alternatives that should be explored?

The committee finds that a comprehensive set of cesium partitioning alternatives was identified in the Savannah River Site's screening procedure, and it recommends that no further effort be expended in alternatives identification at this time. Additional details can be found in Chapter 2 of this report.

Charge 2: Was the procedure used to screen the alternatives technically sound and did its application result in the selection of appropriate preferred alternatives?

The committee finds that the screening procedure was cumbersome, complex, and lacked transparency to document the technical soundness of the evaluations. The selection of alternatives appears to be based primarily on the best judgment of experts using many qualitative factors, an appropriate technique whose merit was obscured by the attempt to quantify the qualitative judgements in voluminous documents. The screening procedure did, however, result in the selection of four potentially appropriate processing options. Additional details can be found in Chapter 2.

Charge 3: Are there significant barriers to the implementation of any of the preferred alternatives, taking into account their state of development and their ability to be integrated into the existing HLW processing system?

The committee finds that there are potential barriers to implementation of all of the alternative processing options. The committee recommends that Savannah River proceed with a carefully planned and managed research and development (R&D) program for three of the four alternative processing options (small tank precipitation using TPB, crystalline silicotitanate ion exchange, and caustic side solvent extraction, each including monosodium titanate processing for removing strontium and actinides) until enough information is available to make a more defensible and transparent downselection decision. The budget for this R&D should be small relative to the total cost of the processing program, but this investment will be invaluable to overcoming many of the present uncertainties discussed in this report. Recommendations on the elements of this R&D program are provided in Chapters 3-6. Additionally, the committee recommends that DOE hold good faith discussions with regulators to determine if the direct grout option is feasible should all of the other processing options, including those that are potentially viable but had been previously discarded, prove to be technically or economically impractical. Additional details on this recommendation can be found in Chapter 7.

The committee makes additional recommendations to overcome more "global" barriers for implementing any of these processing options in Chapter 8 of this report. In particular, the committee recommends that the Savannah River Site implement a more fully integrated systems engineering approach for processing HLW salt solutions. This approach involves the possibility of tailoring of HLW salt processing to the contents of individual waste tanks, as well as possible changes to the order of radionuclide removal. Given the large variability in radionuclide content between tanks, it is not clear to the committee that SRS has adequately addressed whether all tanks should be subjected to the same processing operations. Analysis of this issue should be predicated on possible reduction in size, cost, and design requirements for selected processing options. The committee also recommends that DOE charter external technical review and oversight groups to guide, evaluate, and provide direction to the R&D needed to support this program and to the managers of the program responsible for making decisions. DOE Order 435.1 places responsibility for the salt processing program with SRS. Careful coordination will be required between DOE Headquarters staff and SRS managers to implement these recommendations.

Charge 4: Are the planned research and development activities, including pilot-scale testing, adequate to support implementation of a single preferred alternative?

The answer to this charge's question, at the time the committee completed its information gathering, is "NO!" The committee makes numerous recommendations in this report on the elements of an effective R&D program to address the remaining scientific and technical uncertainties with each of the four alternative processing options. R&D plans were being made by DOE as the committee finalized this report, so the committee has not had the opportunity to review and evaluate the final plan and its progress. Detailed recommendations on these R&D elements can be found in Chapters 3 through 7 of this report.

1

Introduction

At the request of the U.S. Department of Energy (DOE), the National Research Council (NRC) empanelled a committee (see Appendix A for membership) to provide an independent technical review of alternatives to the discontinued in-tank precipitation (ITP) process for treating the high-level radioactive waste (HLW) stored in tanks at the Savannah River Site (SRS).[1] In this chapter the committee presents background information on the high-level waste program at SRS, the origin of the cesium removal problem, processes for cesium removal, and preliminary recommendations from the committee's interim report on this study (National Research Council, 1999b; see Appendix B). Since the interim report was issued, the committee has sought clearer, more definitive answers to the questions and concerns it identified in that report.

Under the statement of task for this study, the committee reviewed the DOE work to identify alternatives for separating cesium from high-level waste at the Savannah River site. This review addressed the following four charges from DOE:

- Was an appropriately comprehensive set of cesium partitioning alternatives identified and are there other alternatives that should be explored?
- Was the process used to screen the alternatives technically sound and did its application result in the selection of appropriate preferred alternatives?
- Are there significant barriers to the implementation of any of the preferred alternatives, taking into account their state of development and their ability to be integrated into the existing SRS HLW system?

---

[1] Throughout this report the committee uses the term "Savannah River Site" (abbreviated "SRS") in general to identify the site and those working within it. A specific reference will be used when appropriate to address another particular site or group (e.g., Oak Ridge National Laboratory, Westinghouse Savannah River Company, or Savannah River Technology Center).

- Are the planned research and development (R&D) activities, including pilot-scale testing, adequate to support implementation of a single preferred alternative?

BACKGROUND ON THE HIGH-LEVEL WASTE PROGRAM AT SAVANNAH RIVER

During and immediately following the Second World War, the U.S. Government established large industrial complexes at several sites across the United States to develop, manufacture, and test nuclear weapons. One of these complexes was established in 1950 at SRS to produce isotopes, mainly plutonium and tritium, for defense purposes. The site is located adjacent to the Savannah River near the Georgia-South Carolina border and the city of Augusta, Georgia (Figure 1.1), and comprises an area of about 800 square kilometers (~300 square miles).

The SRS is host to an extensive complex of facilities that included fuel and target fabrication plants, nuclear reactors, chemical processing plants, underground storage tanks, and waste processing and immobilization

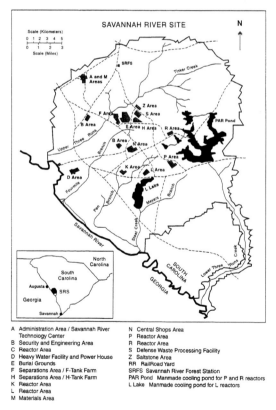

FIGURE 1.1 Location and layout of the Savannah River Site. SOURCE: U.S. Department of Energy.

facilities. Plutonium and tritium were produced by irradiating specially pre-pared metal targets in the nuclear reactors at the site. After irradiation, the targets were transferred to the *F Canyon* or *H Canyon*, where they were processed chemically to recover these radionuclides. This processing re-sulted in the production of large amounts of highly radioactive liquid waste, known as *high-level waste*, that, after treatment with caustic, is being stored in two underground tank farms at the site (in the *F Tank Farm* and *H Tank Farm*). DOE has the responsibility for waste management at SRS and has implemented a program to stabilize this HLW and close the tank farms. A simplified schematic representation of the tank waste processing system at SRS is shown in Figure 1.2.[2] This system comprises the major components; (a) waste concentration and storage, (b) radionuclide immobilization, (c) ex-tended sludge processing, (d) salt processing, and (e) salt disposal.

Waste Concentration and Storage

The high-level waste resulting from operations in the F and H Can-yons is currently being stored in 48 underground carbon-steel tanks[3] in the F and H Tank Farms. The tanks range in size from about 3 million to 5 million liters (750,000 to 1.3 million gallons). The HLW was made alkaline with so-dium hydroxide (NaOH) and formed a caustic sludge before being transferred to the tanks to reduce corrosion of the carbon steel primary containment. Consequently, the waste has a high pH (>14) and a high salt (especially so-dium) content. A summary of the SRS tank wastes is shown in Table 1.1, and Table 1.2 reports key waste components for the individual tanks.[4]

Approximately 400 million liters (100 million gallons) of HLW were produced at SRS since operations began in the 1950s, but this volume has been reduced to about 130 million liters (34 million gallons) by removal of excess water through evaporator processing operations. About 10 percent of

---

[2] The flow charts shown in Figure 1.2 and others throughout this report are schematic and not intended to be quantitative. Since none of the chemical processes discussed is 100 percent efficient, each of the separation steps is shown to leave behind some of the radionuclides that were to be removed. Similarly, chemical separations are never com-pletely specific for a given radionuclide, so the flow charts show that some of each species present is removed in a given operation. Finally, the sequence of operations shown is somewhat arbitrary, corresponding to that proposed by SRS, but other sequences are possible. For example, cesium removal may precede or follow strontium and actinide re-moval, or several separations may be combined into a single step (see Chapter 8).

[3] There are 51 tanks in the F and H Tank Farms; of those, four are process tanks not being used for storage, two have been filled with grout, and one is empty.

[4] The data used to prepare these tables are from the Savannah River Waste Char-acterization System database and were provided to the committee by representatives at Westinghouse Savannah River Company in a specially prepared document (Fowler, 2000). According to SRS, these data represent the site's best-available knowledge about the contents of the HLW tanks. The committee cannot verify the accuracy of these data and notes that some of the data appear to be of questionable quality.

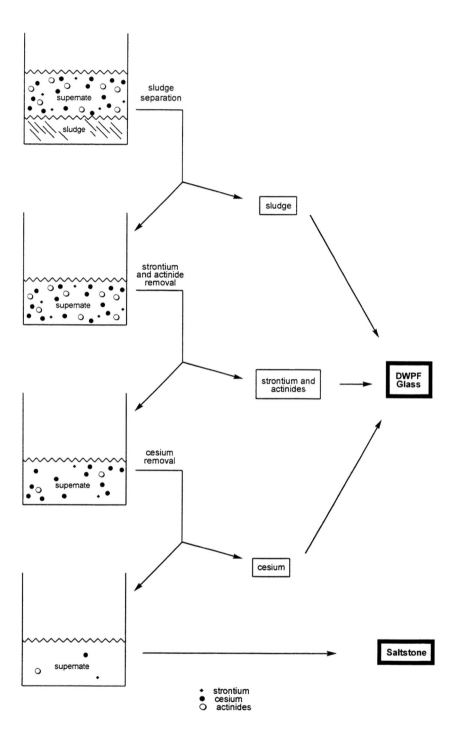

FIGURE 1.2 Schematic processing flow sheet for radionuclide removal from high-level waste at Savannah River.

Table 1.1 Characteristics of High-Level Waste Tank Waste at Savannah River Site

	Sludge	Saltcake	Free Super-nate[b]	Total Soluble	Total
Volume[a] (10^6 L)	9.4	58.4	65.5	na	133
Dry mass (10^6 kg)	2.5	22.4	na	na	na
Cesium-137 (10^6 Ci)	5.9		88.8[e]	95[c]	95[c]
Strontium-90 (10^6 Ci)	98[d]		0.03[e]	0.03	98
Total Alpha (10^6 Ci)	3.6[d]		0.1[e]	0.1	3.7

[a] Both sludge and saltcake contain interstitial supernate, assumed to have the same concentrations of radionuclides as the free supernate.

[b] Does not include the interstial supernate in sludge (5.7×10^6 L) and saltcake (13×10^6 L).

[c] SRS estimates that 90-95% of the cesium in the sludge will dissolve during sludge washing.

[d] SRS estimates that ^{90}Sr and alpha components of sludge will remain insoluble during washing.

[e] Estimates reflect the sum of radionuclides for saltcake and free supernate.

SOURCE: Data from Fowler (2000).

the waste by volume is in the form of a water-insoluble precipitate, or *sludge*, that contains most of the actinides (i.e., uranium as well as transuranic elements) and strontium-90. This sludge was formed by natural settling and by precipitation when NaOH was added to the waste. The remaining waste consists of solid sodium salts (*saltcake*) and an aqueous solution (saturated with sodium salts) called *supernate*. The supernate contains approximately 95 percent of the cesium in the tank waste, as well as minor amounts of actinides. The saltcake, produced by crystallization after the alkaline waste was processed through evaporators to remove excess water, will dissolve when additional water is added during waste processing. The saltcake and sludge contain substantial quantities of supernate within their mass; this interstitial supernate corresponds to about half of the total supernate in the tanks. Altogether, approximately half of the total radioactivity in the tanks resides in insoluble form in the sludge and half is either dissolved or expected to dissolve during processing of the tank waste.

Radionuclide Immobilization

The Defense Waste Processing Facility (DWPF) was constructed to immobilize radioactive waste in borosilicate glass for eventual shipment to and disposal in a geological repository. The glass-making process is referred

TABLE 1.2 Waste Volumes and Principal Radionuclide Contents of High-Level Waste Tanks at the Savannah River Site

Tank Number	Waste Volumes (10^6 L)				Soluble Radionuclides			Sludge
	Total Waste	Sludge	Saltcake	Free Supernate	Alpha (10^3 Ci)	^{90}Sr (10^3 Ci)	^{137}Cs (10^6 Ci)	^{137}Cs (10^6 Ci)
1	1.9	0.03	1.8	0.07	0.03	0.7	2.4	0.05
2	2.1	0.02	2.0	0.001	0.03	0.04	0.8	0.006
3	2.1	0.02	2.0	0.002	0.03	0.04	0.8	0.005
4	1.9	0.5	0.1	1.3	0.09	0.08	3.6	0.29
5	0.1	0.05	0.1	0.0	0.0007	0.005	0.1	0.21
6	1.3	0.1	0.0	1.2	0.07	0.4	0.1	0.25
7	1.4	0.8	0.0	0.6	0.06	0.02	0.9	0.28
8	0.6	0.5	0.0	0.1	0.02	0.02	0.08	0.25
9	2.1	0.02	2.0	0.007	0.03	0.06	0.8	0.006
10	0.7	0.02	0.7	0.0	0.01	0.01	0.06	0.0006
11	1.2	0.5	0.0	0.7	0.06	0.05	0.3	0.32
12	0.7	0.3	0.3	0.0	0.004	0.003	0.09	0.50
13	3.4	0.8	0.0	2.5	0.2	0.0	9.1	0.50
14	0.7	0.1	0.6	0.0	0.01	0.2	0.8	0.01
15	0.8	0.8	0.4	0.0	0.005	0.0	0.0007	0.44
18	1.3	0.2	0.0	1.2	0.07	0.08	0.0003	0.0007
19	1.1	0.03	0.1	1.0	0.06	0.05	0.008	0.00004
21	3.8	0.05	0.0	3.8	0.2	0.0	0.007	0.003
22	4.9	0.08	0.0	4.8	0.3	0.0	0.001	0.007
23	3.6	0.2	0.0	3.4	0.2	0.0	0.0001	0.0005
24	1.1	0.0	0.0	1.1	0.06	0.0	0.007	0.0
25	4.8	0.0	4.2	0.6	2.9	0.05	1.7	0.0
26	2.9	1.1	0.0	1.8	4.8	0.1	0.4	0.01
27	4.8	0.0	1.8	3.1	6.5	0.1	3.5	0.0
28	4.6	0.0	3.9	0.7	2.9	0.05	1.7	0.0
29	4.7	0.0	3.8	0.9	3.2	0.7	4.1	0.0
30	3.6	0.002	0.3	3.4	6.5	22.0	11.0	0.002
31	4.8	0.0	3.8	1.0	3.4	0.0	5.1	0.0
32	4.7	0.7	0.0	4.0	8.4	2.8	6.4	0.6
33	2.0	0.1	0.9	1.0	2.4	0.06	0.04	0.4
34	4.4	0.09	0.8	3.5	7.0	0.0	3.2	0.6
35	3.8	0.2	0.0	3.5	6.9	0.07	5.3	0.54
36	4.7	0.001	4.1	0.6	2.8	0.0	7.9	0.0006
37	4.7	0.0	3.7	1.0	3.4	0.0	6.8	0.0
38	3.9	0.0	3.4	0.5	2.4	0.1	0.1	0.0
39	4.0	0.4	0.0	3.6	7.3	0.1	2.8	0.52
40	1.4	0.5	0.0	0.8	2.3	0.05	0.3	0.03
41	4.7	0.0	4.7	0.07	2.1	0.02	0.7	0.0008
42	4.5	0.05	0.0	4.4	8.4	0.02	1.2	0.02
43	4.5	0.2	0.5	3.9	7.7	0.3	0.22	0.06
44	4.5	0.0	3.7	1.1	3.6	0.0	2.5	0.0
45	4.8	0.0	4.3	0.5	2.7	0.0	1.8	0.0
46	4.8	0.0	1.2	3.3	6.7	0.0	0.8	0.0
47	4.5	0.9	3.3	0.5	3.5	0.02	1.2	0.01
Total	133.2	9.4	58.4	65.5	109.3	28.3	88.8	5.92

Notes: Tanks 16, 17, 20, and 48 through 51 are not listed in this table as they are not used for waste storage. The zero values shown in this table are given as blank entries in Fowler (2000).
SOURCE: Fowler (2000)

to as *vitrification*. This glass is produced by combining the processed HLW (the processing operations are discussed below) with specially formulated glass frit and melting the mixture at about 1150 °C. The molten glass is then poured into cylindrical stainless steel canisters, allowed to cool, and sealed. The DWPF canisters are about 60 centimeters (2 feet) in diameter and about 300 centimeters (10 feet) in length and contain about 1,800 kilograms (4,000 pounds) of glass. About 700 canisters have been produced to date, and SRS estimates that a total of about 6,000 canisters would be produced by 2026, when the tank waste processing program is planned to be completed. These canisters are to be stored at the site until a permanent geological repository is opened and ready to receive them.

Extended Sludge Processing

Extended sludge processing would be used to prepare the sludge portion of the tank waste for processing into glass. The sludge is removed from the tanks by hydraulic slurrying and washed to remove aluminum and soluble salts, both of which can interfere with the glass-making process. The washed sludge is transferred to the DWPF for further processing before being incorporated into glass. Sludge processing, part of the four alternatives discussed in this report, would result in immobilization in glass of nearly all of the strontium and actinides from the tanks.

Salt Processing

Salt processing would be used to remove much of the radionuclides from the HLW salt for eventual vitrification. The salt is to be redissolved and transferred out of the tanks. It would then be mixed with a sorbent to remove any remaining actinides (mainly uranium and plutonium) and strontium. The currently planned sorbent is monosodium titanate (MST). The solution will then be subjected to another (and as-yet undetermined) process to remove cesium. This processing step, along with strontium and actinide removal, are the foci of the present study. The separated actinides, strontium, and cesium would be washed to remove soluble salts and sent to the DWPF for immobilization. Even for the direct grout option, soluble actinides and strontium would be removed by MST treatment, leaving cesium in the aqueous stream.

Salt Disposal

A variety of secondary waste streams are formed during the processing operations described above. Some of these waste streams are recycled back to the tanks, some are recycled within the various processing operations, and yet other wastes are treated and stabilized for burial. Most notably, the "decontaminated" salt supernate (i.e., the solutions remaining after actinide, strontium, and cesium removal) would be disposed of onsite in a waste form known as *saltstone*. The residual solutions are classified as "incidental waste" from the processing of HLW. Saltstone is created by mixing the residual salt solutions with fly ash, slag, and portland cement to create a grout slurry. This slurry is then poured into concrete vaults, where it cures (solidifies) and is eventually covered with soil. The saltstone contains small quantities of some radionuclides (e.g., cesium-135 and -137, having half-lives of about 2 million years and 30 years, respectively), as will be discussed in Chapter 7. The Saltstone Production Facility is permitted by the South Carolina Department of Health and Environmental Control as a waste water treatment facility. The saltstone vaults are designed as a controlled release landfill disposal site. The operating permit limits the average concentrations of radioactive contaminants to the limits specified by the U.S. Nuclear Regulatory Commission for Class A Waste in 10 CFR 61. In the direct grout disposal option, which is discussed below, cesium also would be immobilized in the grout, possibly resulting in increased concentrations of radionuclides up to the limits of Class C waste.

At present, SRS is processing sludge from the tanks to make glass at the DWPF. The current HLW processing schedule calls for the salt solutions to be treated to recover the actinides, strontium, and cesium, beginning about 2008. The 2008 schedule has been proposed to maintain operations at the DWPF and to ensure that there is sufficient space in the tank farms to continue operations at the site.[5] To meet this schedule, however, SRS must develop, test, and implement a process for removing actinides, strontium, and cesium from the radioactive salt in the tanks.

ORIGIN OF THE CESIUM REMOVAL PROBLEM

As noted above, SRS planned to remove actinides, strontium, and cesium from the salt solutions in two processing steps. First, actinides and strontium were to be removed by mixing the salt solutions with MST, resulting in the sorption of actinides and strontium. The product of this reaction could be removed from the salt solutions by filtration for subsequent processing and immobilization. Subsequently, the removal of cesium from the salt solutions would be accomplished by a yet-to-be-chosen process from among precipitation, ion exchange, or solvent extraction processes.

In the late 1970s and the 1980s, SRS developed a process for removing cesium from salt solutions through a precipitation reaction involving sodium tetraphenylborate (NaTPB) and cesium to form cesium TPB (CsTPB):

---

[5] At present, the F and H Tank Farms have about 2.6 million liters (700,000 gallons) of empty space. SRS is reevaluating its long-term tank space needs.

$$Na^+ + TPB^- + Cs^+ \rightleftharpoons Na^+ + CsTPB\downarrow$$

SRS refers to this process as "in-tank precipitation." The NaTPB was to be added directly to a large waste tank to produce a cesium-bearing precipitate. SRS undertook an ITP pilot project in 1983 to demonstrate proof of principle. The process removed cesium from the salt solution, but it also resulted in the generation of flammable benzene from radiolytic reactions and possibly from catalytic reactions with trace metals in the waste.

In September 1995, SRS initiated ITP processing operations in a tank that contained about 1.7 million liters (450,000 gallons) of salt solutions. The operations were halted after about 3 months because of higher-than-expected rates of benzene generation. SRS staff then initiated a research program to develop a better understanding of the mechanisms of benzene generation and release. They also considered possible design changes to handle the benzene during processing operations and catalyst poisoning strategies. In 1996, the Defense Nuclear Facility Safety Board (DFNSB) issued Recommendation 96-1, urging DOE to halt all further testing and to begin an investigative effort to understand the mechanisms of benzene formation and release with the following recommendation (Defense Nuclear Facilities Safety Board, 1996; see Appendix F):

> *The additional investigative effort should include further work to (a) uncover the reason for the apparent decomposition of precipitated TPB in the anomalous experiment, (b) identify the important catalysts that will be encountered in the course of ITP, and develop quantitative understanding of the action of these catalysts, (c) establish, convincingly, the chemical and physical mechanisms that determine how and to what extent benzene is retained in the waste slurry, why it is released during mixing pump operation, and any additional mechanisms that might lead to rapid release of benzene, and (d) affirm the adequacy of existing safety measures or devise such additions as may be needed.*

Investigations by SRS in 1997 uncovered the possible role of metal catalysts in the benzene formation process. SRS concluded, however, that both safety and production requirements could not be met, which led to the suspension of operations altogether in early 1998. At the time of suspension, SRS had spent almost a half billion dollars to develop and implement the ITP process.

IDENTIFICATION OF PROCESSES FOR CESIUM REMOVAL

In March 1998, Westinghouse Savannah River Company (WSRC) formed a systems engineering team to identify alternatives to the ITP process for separating cesium. This team undertook a literature and patent screening procedure to identify currently known processes, followed by a system of analyses by panels of experts to reduce the number of alternative processes to four. The selection procedure is discussed and evaluated in Chapter 2.

Strontium/Actinide Removal by MST

In all four of the final candidate processes for cesium separation, prior removal of strontium and actinides is viewed by SRS as a requisite process. At present, the use of MST is the method of choice. Some technical uncertainties remain to be resolved, of which the major ones are the kinetics of sorption on MST and the amount of titanate acceptable for proper quality of the vitrified waste form. Work to address these concerns and consideration of alternative processes are discussed in Chapter 3.

Tetraphenylborate Precipitation Process

The ITP developed by WSRC removes cesium from HLW supernates by precipitation with tetraphenylborate ion, $[B(C_6H_5)_4]^-$ (TPB). Sodium TPB is a reagent used for analyzing for the potassium ion based on the insolubility of potassium TPB (KTPB). The 200-fold lower solubility of cesium TPB (CsTPB) can provide decontamination factors (DF) from the salt as high as 10^5 to 10^6 and the mixed CsTPB/KTPB precipitate is typically in a form that is easily filtered. On average, SRS HLW contains sodium ions (approximately 5 molar), potassium ions (approximately 0.03 molar), and cesium ions (approximately 0.00025 molar).

HLW treatment, including the removal of cesium-137, involves separation of selected radioactive components and their subsequent immobilization in a borosilicate glass at the DWPF. To prevent organic material from being fed to the DWPF melters, the CsTPB/KTPB precipitate must be treated to remove more than 90 percent of the phenyl (C_6H_5) groups bound to the boron. Thus, a precipitate hydrolysis process (PHP) was developed to hydrolyze the TPB using formic acid in the presence of a copper catalyst. The hydrolysis products are benzene, which is removed by evaporation and incineration, and an aqueous solution containing $^{137}Cs^+$, $B(OH)_3$, and K^+ ions. An attractive feature of TPB is its susceptibility to catalytic decomposition downstream. The present status of ITP and an analysis by this committee of the small tank TPB process is discussed more fully in Chapter 4.

Crystalline Silicotitanate Ion Exchange

Ion exchange has been in commercial use for over 100 years to remove ions from aqueous solutions, for example, to make deionized water. In most applications the separated ions are *eluted* from the ion exchange material, for example, using a dilute acid, the eluted ions are concentrated, and the ion exchanger is reused over and over. Although this technology is well established, ion exchange for cesium removal from high-level waste at SRS and other DOE sites poses challenges. The ion exchange material must withstand both high alkalinity and high radiation fields and must be very selective for cesium in the presence of much higher concentrations of the chemically related sodium and potassium ions. A promising material for use by SRS to remove cesium is crystalline silicotitanate (CST), developed by Sandia National Laboratory and Texas A&M University, based on work performed on amorphous hydrous titanium oxide in the 1960s and 1970s at Sandia. It was developed as an ion exchange medium known as "TAM-5" at Sandia and Texas A&M under the auspices of the DOE Environmental Management's (EM's) Office of Science and Technology. In 1992, UOP of Des Plaines, Illinois, commercialized the material under the names of "IONSIV® IE-910" for the powdered form and "IONSIV® IE-911" for the engineered form.

CST has received considerable attention because of its promise as an ion exchange material for nuclear waste applications. The material has a high selectivity for Cs^+ in salt solutions over a large portion of the pH range from acidic to basic solution, and exhibits high stability to radiation as well. CST is also unusual in that cesium is difficult to remove from the material (i.e., it is nonelutable and the CST cannot be reused). As a result, CST must be incorporated into the HLW stream along with the radionuclides, and the stability of borosilicate glass with higher concentrations of titanium is an issue that must be addressed. This and other aspects of the CST process are described in Chapter 5.

Caustic Side Solvent Extraction

A typical solvent extraction process includes four steps. First, a feed stream is contacted with a solvent that is virtually insoluble in the stream. During this contact, one or more components of the stream transfer to the solvent, while other components do not. The loaded solvent, scrubbed to remove minor contaminants and leaving relatively clean solvent plus the component(s) to be finally recovered, is sent to a stripping operation where the component(s) to be recovered is(are) removed. The stripped solvent may then go to a solvent-recovery step, in which it is cleaned prior to returning to the first step. In such a process, very high removals of extracted components often can be attained.

Solvent extraction has had a long history of successful use in the nuclear industry for such operations as spent fuel reprocessing and plutonium recovery. This history includes long periods of time in which solvents of various organic species have been exposed to high-radiation fields without experiencing catastrophic degradation rates. Solvent extraction operations usually consist of selectively transferring components from an aqueous, acidic stream into the organic stream. A second aqueous stream of somewhat different composition is often used to strip the solvent and concentrate the extract. For the SRS application, the solvent extraction process must remove approximately 99.998 percent of the cesium (a decontamination factor, or DF, of 50,000) from an aqueous, tank-waste feed stream. The raffinate aqueous stream, thus purified of cesium, would be sent to the SRS saltstone facility, and the extract, concentrated in cesium by about an order of magnitude is sent to the DWPF.

In spite of the problems associated with developing an extractant system, which is highly selective for cesium as well as thermally, chemically, and radiologically stable, there are several potential advantages for such a process at SRS. The possible use of one promising system is discussed in Chapter 6.

Direct Disposal in Grout

Direct disposal of the tank waste following removal of strontium and actinides is very similar to the saltstone process that was to have been used to dispose of the salt solutions from ITP operations as low-level incidental waste. Although it is a rather mature technology and has already been demonstrated at the site for less radioactive salt solutions, the degree of retention of cesium may not satisfy regulatory requirements. These and other concerns over this disposal option are addressed in Chapter 7.

RECOMMENDATIONS FROM THE COMMITTEE'S INTERIM REPORT

In October 1999, the NRC Committee on Cesium Processing Alternatives for High-Level Waste at the Savannah River Site issued an interim report (National Research Council, 1999b; see Appendix B) that provided a preliminary assessment of the process by which the final alternatives were selected by the SRS. Although the committee noted in that report that it did not yet have enough information to address fully the committee's stated task, it reached several conclusions relative to that task and offered several recommendations based on the information gathered to that time.

As Task 1, the committee had been asked to evaluate the screening process used by SRS to identify alternatives for cesium removal from the SRS HLW tanks. The committee's preliminary impression was that the screening process did result in the identification of several potentially viable alternative processing options, although it was difficult to determine at that point the thoroughness and objectivity of the process.

The committee's Task 2 was to review the selection of the final four options and their appropriateness to the problem. The committee could provide only a cursory evaluation of the list of 18 alternative processing options developed by the SRS High-Level Waste Salt Disposition Systems Engineering Team. However, it noted that these options included the approaches known to committee members to be useful for processing cesium-bearing alkaline salt solutions, and it concluded that no obvious major processing options had been overlooked in the screening process.

For Task 3, regarding identification of any significant barriers to the implementation of the final four options selected for cesium processing, the committee concluded that any of these four alternative processes probably could be made to work if enough time and funding were devoted to overcoming the remaining scientific, technical, and regulatory hurdles.

In response to Task 4, to comment on the adequacy of the planned research and development activities for the final four alternative processes, the committee observed that "R&D resource allocations for the four alternative processing options have been markedly inequitable." This funding disparity explains in part the different levels of technical maturities of the four processing options, independent of their likelihood of success. The committee also stated that its discussions with SRS staff and contractors gave a strong impression that the primary contractor at SRS, WSRC, failed to pursue R&D on any option for cesium removal but small tank using tetraphenylborate precipitation. The committee also observed that SRS did not even appear to have a well-developed R&D plan for the small tank TPB precipitation and the CST ion exchange options as no written R&D plans for any of these options were available, nor was the committee given any description of a R&D program for these options.

In regard to the "front end" option of strontium and actinide removal by MST, the committee noted that some technical questions needed to be resolved before this process could be implemented successfully at SRS. However, SRS did appear to be conducting R&D to answer these questions. A committee concern was the apparent lack of options identified should MST prove to be inadequate. It is noted that review of the front-end processing was not part of the requested tasks for the committee; however, the process is so related to and interactive with all of the cesium separation options that the committee concluded that review and evaluation of the MST process should be included in its report.

In its interim report (National Research Council, 1999b), the committee made the following recommendations:

1) SRS should pursue vigorously one primary and several backup options for processing the cesium-bearing salt solutions at Savannah River until the remaining technical and regulatory issues are resolved.

2) SRS should continue its efforts to address the remaining technical questions concerning reaction kinetics of the MST process for front-end removal of actinides and strontium from the tank wastes and proceed to pilot-scale testing as soon as possible.

3) SRS should develop and implement a vigorous, well-planned, and adequately funded R&D program to address the remaining scientific hurdles with the small tank TPB precipitation process.

4) A vigorous, well planned, and adequately funded R&D effort should be undertaken to address the remaining scientific hurdles with the CST ion exchange option.

5) A vigorous, well planned, and adequately funded R&D effort should be undertaken to address the remaining scientific and technical hurdles with the caustic side solvent extraction option.

6) SRS and DOE should undertake a vigorous program to determine the regulatory acceptability of the direct grout option through discussions with relevant staff at DOE, the U.S. Nuclear Regulatory Commission, the U.S. Environmental Protection Agency, and the South Carolina Department of Health and Environmental Control.

In fact, these recommendations formed the basis for further information gathering and analyses by the committee during this study, leading to the recommendations presented in Chapter 8.

GENERIC ISSUES

Two distinct categories of generic issues are related to treatment of the stored nuclear wastes at SRS; those issues that are overriding and must be treated independently, and those that are repetitive and must be treated in each of the detailed parallel topics of this report. The generic issues in the first category include the need to analyze (1) the SRS waste treatment process as a complete system, and (2) the MST process as an integral part of all the alternatives under consideration. Those in the second category of generic issues include the need to direct attention to (1) barriers to implementation, (2) R&D requirements, (3) adequate demonstration, (4) conformance to SRS schedules, (5) assurance of a minimum risk of failure, (6) reasonable cost, (7) sensible accommodation to existing facilities, and (8) implementation with the overall system. These topics are not completely independent, but each introduces distinct considerations and subsequent constraints.

Known barriers to implementation vary in nature, including those describable with the following adjectives: regulatory, technical, safety, and economic. Different concerns are associated with the different alternatives; that is, grout faces the most formidable regulatory barriers but is the most mature technically, while solvent extraction has fewer regulatory barriers but is the least mature technically. The TPB and front end strontium and actinide removal MST processes have no particular regulatory barriers. Safety concerns are a necessary aspect of the decision in selection among the alternatives for cesium treatment and must be included in the evaluation of any competing process.

Economic analysis results and cost must also be factors that are considered in the final selection process. The need for R&D is far from uniform across the alternatives; some are almost ready for pilot plant demonstrations, while others require significantly more R&D. Ultimately, prior to a final choice of the process to use, sufficient development needs to be conducted to bring each alternative process to a level of maturity such that an informed comparison and evaluation can be made among all alternatives. Such an analysis must consider whether each alternative could be successfully implemented and what potential barriers might remain to be overcome. The proposed R&D must be focused on the development pathway and include appropriate decision points. Pilot scale demonstrations, including adequate hot operations using actual representative radioactive waste from the tanks, are critically important for process evaluation. Two economic as well as technical issues that will require constant attention as the R&D program proceeds are minimizing the risk of process failure and careful analyses of opportunities to continue use of existing facilities and resources for research and operations. Further, as noted above, no demonstration or accommodation is complete without consideration of the total waste management system. Whereas SRS schedules may be undergoing change at any point in time, it is necessary that proposed evaluation, R&D, and demonstration schedules correspond as well as possible with the time requirements imposed by SRS for implementation of the waste treatment.

2

Screening Procedure

In response to the increased awareness of concerns about the safety and efficacy of the large tank in-tank precipitation (ITP) process selected for the removal of cesium from the tank waste supernate (U.S. General Accounting Office, 1999; Defense Nuclear Facilities Safety Board, 1996), Westinghouse Savannah River Company (WSRC) undertook a study to identify and evaluate alternatives to ITP. The Savannah River Site (SRS) and WSRC established a procedure[1] whereby the cesium separations literature and patents were searched, along with evaluation by panels of experts of cesium removal processes, to select alternative options to replace the current, unacceptable ITP process for high-level waste (HLW) salt removal, treatment, and disposal.

BACKGROUND

The procedure was initiated by a comprehensive literature and patent search. Subsequently, through what WSRC called "Phases I, II, III, and IV" (Figure 2.1), 144 cesium removal process alternatives were identified and evaluated by panels of experts in a sequence of steps that reduced the number to four and finally to one recommended alternative and one backup technology. The selection procedure that identified the 144 process alternatives and subsequently winnowed down the number of alternatives to one recommended and one backup process was based on the expert judgment of representatives from various U.S. Department of Energy (DOE) laboratories and consultants. Although the interim narrowing to one main process and a

---

[1] The term *procedure* is used herein to describe the methodology of evaluation, review, selection, ranking, etc., of operations related to the separation of cesium and other elements and the production of waste forms; the term *process* is used herein to refer to the operations, usually chemical engineering related, of carrying out separations and production of waste forms.

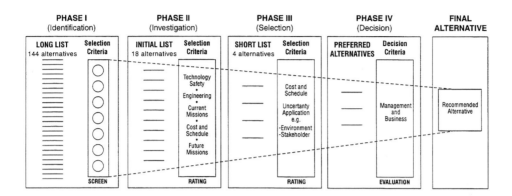

FIGURE 2.1 Schematic outline of the screening procedure used to identify a recommended alternative for cesium removal. SOURCE: Westinghouse Savannah River Company.

backup has been conducted, the committee understands that the current plans by WSRC appear to include an evaluation of several other highly ranked process alternatives.

Literature and Patent Search

The initial literature and patent search, conducted by WSRC (Poirier, Hunt, and Carlson, 1998), interrogated 11 large databases of scientific, engineering, and patent information:

- Chemical Abstract Services
- National Technical Information System (NTIS)
- American Geological Institute's GeoRef Database
- Nuclear Science Abstract
- Engineering Information, Inc.
- Inside Conferences
- Energy Index
- Information Services in Physics, Electronics, and Computing (INSPEC)
- Analytical Abstracts
- World Patents
- The U.S. Patent and Trademark Office Homepage

These databases were integrated using seven selected key words containing various phrases related to cesium separations:

- cesium removal
- cesium separation
- cesium extraction
- cesium precipitation
- cesium ion exchange
- cesium filtration
- cesium concentration

In conducting the search, WSRC focused on identification of separation methods as described by the chemistry of the processes. The resulting assemblage of more than 1,700 citations was grouped into 16 basic process technologies that encompass about 440 varieties of these technologies (Table 2.1).

A ranking of the 16 process technologies uncovered by the literature and patent search revealed that ion exchange, precipitation, adsorption, extraction, filtration, and biological process were the most prominent technologies (over 95 percent) found in terms of numbers of varieties. The largest number of references to varieties of ion exchange was found for crystalline sodium titanate, resorcinol formaldehyde, hexacyanoferrates, and Duolites. Prominent precipitation processes identified by the search included the use

TABLE 2.1 Results of the Literature and Patent Search

1) Adsorption (50 different varieties)
2) Biological Methods (9 different varieties)
3) Centrifugation (1 variety)
4) Chromatography (2 different varieties)
5) Crystallization (1 variety)
6) Electrochemical (5 different varieties)
7) Evaporation (1 variety)
8) Extraction (14 different varieties)
9) Filtration (10 different varieties)
10) Flotation (1 variety)
11) Ion Exchange (285 different varieties)
12) Leaching (1 variety)
13) Magnetic (1 variety)
14) Nanofiltration–Complexation (1 variety)
15) Precipitation (59 different varieties)
16) Pyrochemical (2 different varieties)

NOTE: Over 1,700 literature and patent references were identified and grouped into 16 "process technologies."
SOURCE: Poirier, Hunt, and Carlson (1998); Poirier (1998).

of tetraphenylborate and ferrocyanides/ferrates. Clays, zeolites, and ferro-cyanides/ferrates were the most numerous adsorption chemicals cited in the literature. The topic of solvent extraction yielded extensive references to the use of crown ethers and calixarenes that have not been used at a plant-size scale. This search also yielded a reference to dicarbollide that has been used at a plant-size scale.

Initial Selection of Process Alternatives (Phase I)

The WSRC procedure for identifying and evaluating cesium removal processes throughout the rest of this procedure was guided by a list of critical needs and boundary conditions and constraints that the alternatives must meet (Piccolo, 1999, p. 3) (Table 2.2).

In selecting the initial list of cesium removal alternatives (designated by WSRC as Phase I), a team of experts (Savannah River Site High Level Waste Salt Disposition Systems Engineering Team, 1998a, 1998e) began by

TABLE 2.2 Minimum Critical Needs and Boundary Conditions and Con-straints

---

Critical Needs
- Meet all applicable safety criteria for protection of personnel
- Meet all applicable environmental regulations
- All waste must go to final disposition forms
- Meet federal facility agreements and site treatment plan regulatory commitments
- Accommodate other Savannah River Site (SRS) missions and asso-ciated schedules
- Meet all applicable final disposal product quality requirements
- Meet all applicable waste acceptance criteria

Boundary Conditions and Constraints
- Functions and requirements (preliminary design input)
- Safety of the process
- Impact on high-level waste final waste form disposition
- Programmatic/technical risk
- Cost (project and life cycle)/schedule
- Regulatory/safety/permit acceptability
- Operational complexity
- Ability to support current/planned/future SRS missions and schedules
- Maximum tank farm space kept available
- Use of existing facilities
- Constructability
- Reliability, availability, maintainability, and inspectability

---

SOURCE: Piccolo (1999, p. 3).

grouping the alternative processes into 11 categories.[2] Distributed within these were 144 specific processes[3] (the 'Long List') (Table 2.3). It should be noted, however, that these 11 categories do not necessarily correspond to the processes identified in the initial literature compilation.

TABLE 2.3 Initial Selection of Cesium Removal Process Alternatives (Phase I—'Long List')

1) Crystallization—separation of cesium from non-radioactive salts by fractional crystallization [6 alternatives: 6 accepted and combined into 3, 0 rejected]

2) Electrochemical—processes which achieve separation/destruction of different ionic components in the system [5 alternatives: 5 accepted and combined into 2, 0 rejected]

3) Elutable Ion Exchange—separation of cesium from HLW salt by re-generable ion exchange [17 alternatives: 13 accepted and combined into 3, plus 1 accepted and combined under Non-Elutable Ion Exchange, 4 rejected]

4) Non-Elutable Ion Exchange—separation of cesium from HLW salt by non-regenerable ion exchange [31 alternatives: 25 accepted and combined into 7, plus 2 accepted and combined under Elutable Ion Exchange, 6 rejected]

5) Geological—alternatives more dependent on geology than process-ing [3 alternatives: 0 accepted, 3 rejected]

6) Inorganic Precipitation—separation of cesium by addition of an inor-ganic precipitant [4 alternatives: 0 accepted, 4 rejected]

7) Organic Precipitation/Modify In-Tank Precipitation (ITP)—separation of cesium by addition of an organic precipitant with extensive use of the existing ITP facility [29 alternatives: 21 accepted and combined into 4, plus 4 accepted and combined under Organic Precipitation /New Process, 8 rejected]

8) Organic Precipitation/New Process—separation of cesium using a fa-cility substantially different from the existing ITP facility [17 alterna-tives: 10 accepted and combined into 1, plus 2 accepted and com-bined under Elutable Ion Exchange, 7 rejected]

9) Solvent Extraction—use of a solvent for separating cesium based on either an alkaline or acidic feed stream [6 alternatives: 5 accepted and combined into 2, 1 rejected]

---

[2] These included such options as geologic disposal that were less dependent on separations processing than others.

[3] Although the references cited earlier listed 126 specific processes, the committee was notified on March 9, 2000, by R. Jones, WSRC, that "The activity to create the cesium removal 'Long List' of alternatives began during Phase I and continued throughout Phase III." After publication of the referenced documents concerning Phase I, an additional 18 specific process were identified and added to the list.

Table 2.3 continued

10) Vitrification—disposition of the salt by vitrifying it either in Defense Waste Processing Facility (DWPF) or using new equipment or facilities [8 alternatives: 2 accepted, 6 rejected]
11) Miscellaneous—approaches not covered by the other categories [18 alternatives, including 1 new hybrid: 5 accepted including 1 hybrid, 13 rejected]

NOTE: 144 *Pro Forma* (suggestions, ideas, and concepts) alternatives, designated as the 'Long List', including hybrid alternatives, were grouped based on input from SRS employees, historical reviews, literature and patent search, formal brainstorming, and early informal results from independent subject matter experts. *Pro Forma* alternatives were grouped into 11 "technical categories" (screening criteria for acceptance, rejection, hybridizing, etc., included scientific maturity, engineering maturity, potential for implementation, safety, licensability, response to DNFSB Recommendation 96-1, and feasibility of permitting the final waste form). Eighty-three alternatives were accepted and combined into 19 alternatives and 9 'stand-alones,' for a total of 28; 52 alternatives were rejected.

SOURCE: Savannah River Site High Level Waste Salt Disposition Systems Engineering Team (1998b).

The 144 process alternatives on the 'Long List' were apparently obtained from sources other than directly from the literature search, although the WSRC team received a review of the search and identification of technologies for potential inclusions in the evaluation (Poirier, 1998). Each process input to the procedure was designated as a *pro forma* and was reviewed by the Savannah River Site High Level Waste Salt Disposition Systems Engineering Team (1998b) for relevance, even in the absence of process details. The process outlines, required functions, and corresponding criteria for these candidate processes were submitted and accepted in the usually terse pro forma format.

Once the categories of processes that were considered to be viable for the required mission had been identified, the list of specific candidate processes was narrowed to 83 by formally rejecting 52 of the initial 'Long List.' The reduction of 144 processes to 83 alternatives was largely accomplished by rejecting those that failed to provide a satisfactory end state for the high-level waste, those that had an inadequate scientific base, or those that had obvious and detrimental safety implications. The list of 83 was further reduced by combining some of the process variations and those pro formas that were later submitted (Savannah River Site High Level Waste Salt Disposition Systems Engineering Team, 1998e) into a group of 28 alternatives. This group was reduced to 18 (Table 2.4) by combinations, modifications, or hybrids of the 28 alternatives using screening criteria similar to those mentioned in the note to Table 2.3.

The screening criteria used for this part of the selection process included scientific and engineering maturity, implementation feasibility, safety and licensability, response to the Defense Nuclear Facilities Safety Board (DNFSB) Recommendation 96-1 (1996), and feasibility of obtaining the required permits for disposal of the final waste forms (Savannah River Site High Level Waste Salt Disposition Systems Engineering Team, 1998c).

The reduction of the 28 process alternatives to 18 (the 'Initial List') was achieved by ranking, in each of the process categories, the 28 alternatives and generally, but not always, retaining the uppermost ranked processes (Savannah River Site High Level Waste Salt Disposition Systems Engineering Team, 1998a) (Table 2.4). The ranking was determined on the basis of process robustness, technical maturity, and potential for implementation. Although these terms seem somewhat subjective, WSRC provided numerical guidelines throughout the procedure to quantify the evaluations into ratings. Of the 18 alternatives, 2 solvent extraction processes, 1 selective crystallization process, 2 vitrification processes, 1 grout disposal process, 6 ion exchange processes (of which 3 use crystalline silicotitanate), 1 electrochemical process, and 5 tetraphenylborate processes remained to be evaluated in more detail.

Reduction of Process Alternatives to Four (Phase II)

During Phase II of the screening procedure, the reduction in the number of alternatives from 18 processes to 4 (the 'Short List') was accompanied by an increase in the detailed examination of each of the remaining

TABLE 2.4 Initial Selection of Cesium Removal Process Alternatives (Phase I—'Initial List')

---
1) Fractional Crystallization–DWPF Vitrification
2) Electrochemical Separation and Destruction–DWPF Vitrification
3) Elutable Ion Exchange–DWPF Vitrification
4) Potassium Removal followed by TPB Precipitation
5) Acid Side Ion Exchange–DWPF Vitrification
6) Crystalline Silicotitanate (CST) Ion Exchange–DWPF Vitrification
7) Crystalline Silicotitanate (CST) Ion Exchange–New Facility Vitrification
8) Zeolite Ion Exchange–DWPF Vitrification
9) Crystalline Silicotitanate (CST) Ion Exchange–Ceramic Waste Form
10) Reduced Temperature ITP
11) Catalyst Removal ITP
12) ITP with Enhanced Safety Features
13) Small Tank TPB Precipitation
14) Caustic Side Solvent Extraction–DWPF Vitrification
15) Acid Side Solvent Extraction–DWPF Vitrification
16) Direct Vitrification
17) Supernate Separation–DWPF Vitrification
18) Direct Disposal as Grout

SOURCE: Savannah River Site High Level Waste Salt Disposition Systems Engineering Team (1998b).

options (Savannah River Site High Level Waste Salt Disposition Systems Engineering Team, 1999b and 1998f) (Table 2.5). The substantive steps used to evaluate the initial list of 18 candidate process alternatives included flow sheet analysis, risk analysis, preliminary life cycle cost estimates, reexamination of evaluation criteria, and sensitivity analysis.

TABLE 2.5 Reduction of Process Alternatives to Four (Phase II—'Short List')

1) Small Tank TPB Precipitation
2) Caustic Side Solvent Extraction–DWPF Vitrification
3) Crystalline Silicotitanate (CST) Ion Exchange–DWPF Vitrification
4) Direct Disposal as Grout

NOTE: The Short List was based on information from a preliminary risk assessment, off-site trips, preliminary life cycle cost estimates, and flow sheet analysis. Each alternative was evaluated (i.e., scored) against the following weighted evaluation criteria—technology, current mission interfaces, future mission interfaces, regulatory/integrated safety management system/environmental, engineering (design), and cost/schedule.

SOURCE: Savannah River Site High Level Waste Salt Disposition Systems Engineering Team (1999a,b).

Process flow sheet calculations and the assumptions and bases for the models used to estimate the flow sheet parameters were described in a study that used a preliminary outline of the processes (Westinghouse Savannah River Company, 1998a). The high-level waste (HLW) feed to each of the process models was averaged from all of the tanks, and the end state was assumed to be Defense Waste Processing Facility (DWPF) glass[4] and the saltstone grouting process product. The base case was the ITP process yielding saltstone grout and DWPF glass and was described in some detail.

A preliminary "risk assessment" with adjusted risk values (Savannah River Site High Level Waste Salt Disposition Systems Engineering Team, 1998g), using a general guide to the procedure for the assessment (Savannah River Site High Level Waste Salt Disposition Systems Engineering Team, 1998d) was issued for the 18 process alternatives. The procedure employed by WSRC appeared to be a quantification of potential risks, both technological and regarding implementation, to which were assigned likelihood values and numeric consequence estimates. Numeric values were assigned to each part of the assessment and to the final scores for each candidate process. Where members of the evaluation team identified high risks, mitigation of the risks was evaluated and a modified numerical risk and consequence ranking assigned. For each of the 18 'Initial List' categories, WSRC identified what it called the "highest risks." Those categories having the most high risks were the crystalline silicotitanate process, acid side solvent extraction, small tank tetraphenylborate (TPB), direct grouting, and fractional crystallization. Based on the total number of identified risks of all types, the proc-

---

[4] In some cases, ceramic or special glasses were assumed as the end state.

esses having the most included small tank TPB precipitation, fractional crystallization, direct grouting, catalyst removal in-tank precipitation, and crystalline silicotitanate.

Presentations by WSRC personnel (Piccolo, 1999, p. 8; see also Savannah River Site High Level Waste Salt Disposition Systems Engineering Team, 1999a) indicated that three factors—weighted process evaluation criteria, the preliminary process risk compilation, and flow sheet analysis—were used by the team in deciding which candidate processes to delete from further consideration. On the basis of the scores for technology, four processes were selected to represent the final alternatives, namely direct grouting (score = 22.54), small tank tetraphenylborate precipitation (19.32), zeolite ion exchange (18.86), and crystalline silicotitanate ion exchange (18.86). If only process engineering (construction, operations, reliability, availability, maintainability, and inspectability) weighted evaluation scores are used, caustic side solvent extraction (15.50), acid side solvent extraction (14.25), and direct vitrification (14.00) become the finalists. However, total weighted evaluation scores for all of the final four are probably not significantly different (direct grouting–83.86; small tank tetraphenylborate precipitation–72.88; crystalline silicotitanate ion exchange–69.16; and caustic side solvent extraction––69.14). Zeolite ion exchange received a total weighted evaluation score of 70.65 that would have rated it the number three candidate, but it was rejected in favor of the flexibility of crystalline silicotitanate ion exchange with DWPF vitrification. The direct grout process was excluded from further consideration for non-technical reasons such as the time required to provide for public approval, and regulatory approval (Savannah River Site High Level Waste Salt Disposition Systems Engineering Team, 1999b, p. 40).

Selection of Recommended Process Alternative (Phases III and IV)

The final stage of the procedure (Phases III and IV) for evaluation of processes for removal of cesium resulted in a recommended alternative process (small tank tetraphenylborate) and one back-up process (crystalline silicotitanate ion exchange) (Savannah River Site High Level Waste Salt Disposition Systems Engineering Team, 1998h, 1999a, and 1999b) (Table 2.6). Included in this stage were a more detailed life cycle cost estimate and an uncertainty evaluation to define programmatic risks, and several iterations with other evaluation criteria.

ANALYSIS AND FINDINGS

Based on its review of the documents provided and presentations by and discussions with SRS and its contractors, the committee made the following observations about the screening procedure.

TABLE 2.6 Phases III and IV and Decision Phase

Results
- Recommended Alternative—Small Tank Tetraphenylborate Precipitation
- Backup Technology—Crystalline Silicotitanate (CST) Non-Elutable Ion Exchange

Evaluation Criteria
- technology (scientific and engineering) maturity
- risk management
- worker, public, and environmental safety
- expert judgement
- engineering design
- process simplicity
- impact and interfaces on current and future Savannah River Site missions and program (Defense Waste Processing Facility, saltstone facility, solid waste, tank farm)
- regulatory/permitting
- cost (schedule, life cycle, repository)

SOURCE: Savannah River Site High Level Waste Salt Disposition Systems Engineering Team (1998g, 1999b).

Literature and Patent Search

The initial literature search for process alternatives to separate cesium from the highly alkaline supernate was documented. It appears that a sufficient sample of the pertinent chemistries for the separation of cesium from tank waste supernate is represented. The committee did not receive a summary or analysis of the large amount of written material that resulted from this search procedure, nor was it made aware that such an analysis was available. The committee concluded that, while the scope of the chemistries that were found in the search appeared adequate, it is not clear that the extent of the application of technologies was equally well ascertained. It is unlikely that a common set of key words would satisfy the retrieval of technologies and chemistries from all of the data bases that were searched. Further, some chemistries that have been used on a plant-size scale resulted in only a few references (e.g., cobalt dicarbollide). That circumstance may have been due to the proprietary nature of some information or incomplete information on work that preceded computer archiving. A more comprehensive search may have revealed more of the pertinent information.

Initial Selection of Process Alternatives (Phase I)

The committee concluded that the procedures used to arrive at the 'Initial List' of 18 from a group of 144 processes in the 'Long List' are not obvious, having been obscured in the many volumes of documents containing individual comments and evaluations of members of the expert teams. The value of the extensive global literature and patent search also is not clear, based on the fact that all but one of the technologies (bioremediation, subsequently rejected as not being feasible) were identified by the expert judgment of the Savannah River Site High Level Waste Salt Disposition Systems Engineering Team and its consultants during this part of the procedure. Inclusion of the previously favored processes (the large tank ITP or a variation of it) was maintained throughout the screening procedure; however, the committee was not aware of any additional, substantive information to warrant such retention that might have appeared subsequent to the DNFSB recommendation. It is not clear from the documents provided to the committee how the results of the literature and patent search were incorporated into this step or the subsequent steps of the screening procedure.

The committee concluded that the generation of 18 process alternatives cannot be described as a transparent procedure. Several important separations processes for cesium (e.g., use of cyanoferrates and carbollide extractants) were omitted from the 'Long List'; an internal WSRC note from the leader of the literature and patent search recommended consideration of at least one of them—hexacyanoferrate (Poirier, 1998, p. 2). However, the committee also concluded that the 'Initial List' of 18 alternatives contains processes that probably could be successful in meeting the general objectives of the process within many of the constraints that WSRC has enumerated. Elsewhere in this report, the committee calls attention to the fact that successful application of such separation processes must be based on a rigorous and disciplined program of research and development (R&D).

A comparison between the major chemical processes identified in the literature search (discussed earlier) and the 'Long List' of 144 processes, from which an 'Initial List' of 18 was derived shows significant differences. One of the reasons for these differences was that the literature search was focused on the chemical nature of the separations and primarily identified the central chemical on which the separations rests. The topical array leading ultimately to the final 18 processes does not allow such a ready comparison since it is organized by process (e.g., unit operation) and is often devoid of specific identification of the underlying chemistry.

The 'Long List' included some process alternatives that were obviously not pertinent (e.g., geologic disposal). Further, some of the categories into which the 144 processes were divided were not sought during the literature search (e.g., vitrification); on the other hand, some alternatives (e.g., electrochemical) were included in the 'Long List' of 144 processes even though their presence in the global literature search was trivial. Finally, and most importantly, some of the processes identified in the literature search as extensively described and even applied on a plant scale were not present in the 'Long List'. The committee found no information that identified the reasons for these differences.

The explication of the 18 processes was characterized by WSRC to the committee as truncated. Because of time constraints, the unavailability and uncertainty of some detailed information, especially on the scientific bases for some of the processes, reduced the data base that supported the evaluation.

Reduction of Process Alternatives to Four (Phase II)

The results of the flow sheet analyses are given in terms of rates of use of resources, production of products, or environmental releases. The cited reference (Westinghouse Savannah River Company, 1998a) did not provide conclusions drawn from the flow sheet analyses and assumptions that accompanied the models. Further, the use of monosodium titanate (MST) as a front-end step for many of the alternatives, including the final four processes, apparently was not explicitly assessed in a manner comparable to that used for the alternative processes or integrated with the alternatives. However, the use of titanium in reagents posed questions about the equivalence of titanium in MST and crystalline silicotitanate (CST) in glass chemistry. There appear to be no data to clarify the matter.

The flow sheet calculations for the narrowing of the alternatives from 18 to 4 were claimed to be accurate to ±25 percent, an optimistic claim considering the nature of some of the assumptions used. Neither the flow sheet analysis nor the risk assessment documents provided conclusions that related to the selection of the final four alternatives. Assumptions for the semi-quantitative flow sheets were often invoked and thus indicated a significant absence of reliable and comprehensive data for many of the process steps. Assumptions that appear to be somewhat speculative include (a) reactions proceed to completion, (b) reactions rates for monosodium titanate reach equilibrium in 24 hours for uranium and plutonium (not well known or documented) based on an assumed analogy with the rates for strontium, (c) activity coefficients for tetraphenylborate reactions were calculated but do not seem to be based on experimental data, and (d) concentrations of catalysts, such as copper and palladium, are part of rate equations. In short, the identification of assumptions and uncertainties needed by the expert evaluators provide a general picture of the status of the knowledge for the processes under consideration for alternatives.

The committee notes that this status apparently was not the basis of the risk assessment procedure used by WSRC. Risk assessment is normally used for analysis of the impact of waste disposal or other fuel cycle operations, and its role in this evaluation process was not clear to the committee other than as a guideline for expert judgment. The starting assumptions for the risk assessment were not comparable for each the alternatives (S.F. Piccolo, WSRC, personal communication to committee, November 21, 1999). Unfortunately, the relation between these uncertainties, including those that were brought out during the presentations to the committee, and the final decision matrixes that reduced the number of alternatives to the final four, were difficult to understand.

The committee, through its knowledge of the general literature, as well as from the presentations and the documents provided by WSRC, made an effort to understand how the final four processes were selected and to understand the overall procedure. Although the procedure did not appear to support the processes ultimately selected, this may have been due to the difficulty in understanding and following the details of the procedure. The committee concluded that the steps and procedures for avoiding the potential for less-than-objective analyses were not adequately covered in documents describing the 'qualitative' risk assessment processes used in this phase of the screening process. The selection of the final four alternative processes on the basis of the numeric criteria had little bearing on the extensive literature developed for the analyses. In short, a logical trail that allowed an objective evaluation of the processes used for the selection of the final four candidates in the 'Short List' was not evident. The claim that WSRC has selected the final four alternatives in an obviously objective manner does not seem to be sustainable by the information provided to the committee. The committee concluded, however, that on a technical basis each of the four process alternatives, namely small tank tetraphenylborate precipitation, crystalline silicotitanate ion exchange, direct grout, and caustic side solvent extraction, could be implemented with modest risk if a viable R&D program is carefully planned and successfully executed to address the uncertainties for each. Such a program for each of the four 'Short List' process is discussed in Chapters 4 through 7 of this report.

Selection of Recommended Process Alternative (Phases III and IV)

The relatively high score assigned to the technology rating (science and engineering maturity and process simplicity) for small tank TPB precipitation (78, as compared with the score for the other three process alternatives—49, 58, and 86) and the low score for caustic side solvent extraction (49) in this last step of the screening procedure that narrowed the alternatives to one or two, is at variance with responses provided to the committee during presentations by WSRC personnel. Finally, the technology scoring results appear to be largely insensitive to uncertainties and variations among the candidate processes for process simplicity and engineering maturity. The Savannah River Site High Level Waste Salt Disposition Systems Engineering Team (1999b) introduced the business perspective evaluation, including the entire DOE complex, to arrive at the selection of the small tank ITP as the recommended process, and crystalline silicotitanate in a back-up role. This portion of the procedure was difficult to evaluate.

GENERAL CONCLUSIONS

1) The procedure for identifying alternative processes for separating cesium from HLW was cumbersome, complex, and lacked transparency. The committee concluded that the "winnowing" procedure was sufficiently opaque as presented in the voluminous documentation as to defy ready evaluation of objectivity and completeness. A previous study by a committee of the National Research Council (1998) of systems engineering as applied at the Hanford Site in Washington made a similar observation about the importance of simplifying documentation to improve clarity for reviewers and other interested parties. Nevertheless, the procedure uncovered and, to some extent carried forward, most of the processes that the committee could envision would provide suitable alternatives for cesium removal. There are some alternative processes and varieties not carried forward in the selection procedure because of the emphasis on cesium separations chemistry, but the committee concluded that they do not represent alternatives having a significantly greater likelihood of success or that could be developed much more rapidly than the processes that were selected.

2) The mode of an objective selection of the final process and its backup is not obvious to the committee. The committee concluded that the final phase of the selection process resulting in small tank TPB as the remaining candidate could not have considered the deficiencies in the process brought out during the presentations to the committee. These deficiencies will be discussed further in Chapter 4, but knowledge of them during the selection procedure might have ameliorated the final result.

3) The procedure used to define and narrow the alternative processes for cesium separations raised further concerns:

a) While the numeric (quantified) approach to evaluation seems to have been evenly applied, it leaves an impression of more precision, accuracy, and objectivity than warranted by a procedure depending primarily on expert judgment.

b) More timely attention in the screening procedure to uncovering potential process risks and their scientific and technical uncertainties than was evident for at least the final four cesium separation process alternatives could have lead to recommendations for the R&D necessary to bring each to a point where more rigorous evaluation and final selection could be accomplished. Resolution of important uncertainties that contribute to the technical risk should be identified early in the screening procedure to decrease the need for extensive and impenetrable documentation to support selection decisions for process alternatives.

c) SRS and its contractors appeared to provide inadequate attention in the screening procedure to the role of the cesium separations process selection on the entire HLW system, particularly the steps preceding cesium separations. Evaluation of the explicit interfaces of unit operations (e.g., strontium and plutonium removal) on the risks associated with candidate processes for cesium removal appeared to be incomplete. The committee concluded that a more disciplined systems engineering approach to the entire high-level waste operations at SRS could have clarified many of the issues that remain unresolved (discussed further in Chapter 8 of this report). Further, the impact of qualitative criteria derived from such evaluation factors as perceived future missions at SRS (e.g., weapons-grade plutonium disposition; see National Research Council, 1999a) and site limitations (e.g., availability of tank space) was not clarified in terms of the effect on the selection procedure outcomes. The committee concluded that the overall quality of the outcome of the procedure to select alternatives to the in-tank precipitation was not commensurate with the effort expended to carry it out.

RECOMMENDATIONS

1) SRS should proceed with an R&D program for the four final processes selected unless important barriers arise or until enough information on uncertainties is available to conduct a rigorous but more visible basis for selection.

2) When using qualitative expert judgment, one should not rely on tools such as the numeric (quantified) evaluation procedures that tend to give the false impression of precision, accuracy, and objectivity.

3) In response to the committee task, "Was an appropriately comprehensive set of cesium partitioning alternatives identified and are there other alternatives that should be explored?", the committee recommends that no further effort be expended at this time in alternative identification.

4) In response to the committee task, "Was the process used to screen the alternatives technically sound and did its application result in the selection of appropriate preferred alternatives?", as noted previously the committee concludes that the screening procedure was cumbersome, complex, and lacked transparency to document the technical soundness of an evaluation and selection of appropriate preferred alternatives based primarily on the best judgment of experts using many qualitative factors. The committee recommends that future such evaluations, depending on expert judgment, be documented in a clear, easily understandable and traceable manner to allow for viable reviews. Although the screening procedure did result in the selection in the 'Short List' of what the committee believes are at least four appropriate preferred alternatives, further reduction of the alternatives will have to await completion of adequate R&D on each.

3

Strontium and Actinide Removal

The removal of strontium (Sr) and actinides (especially plutonium [Pu] and neptunium [Np]) is an important step in the salt processing flowsheet at the Savannah River Site (SRS). As presently envisaged, strontium and actinides will be removed from the salt solutions in all four of the cesium processing options discussed in this report—small tank tetraphenylborate (TPB), solvent extraction, ion exchange, and direct grout. Because of its position near the beginning of the processing flowsheet (Figure 1.2), the strontium and actinide removal step is referred to as "front-end" processing. As noted in Chapter 8, however, this processing step does not necessarily need to be performed prior to removal of cesium. In fact, there may be advantages to performing this step later in the processing sequence. This chapter provides a review of this front-end process and a discussion of the remaining technical uncertainties.

Most of the information presented in this chapter was collected at the committee's two information-gathering meetings and from written responses to committee questions outside of those meetings (Westinghouse Savannah River Company, 1999a,c; Jones, 1999a,b; Jones, 2000a,b). Unless otherwise noted, the information used in the discussions in this chapter are taken from these references.

BASELINE APPROACH

The "baseline" approach for this processing incorporates the use of monosodium titanate (MST), $NaTi_2O_5H$, an amorphous solid consisting of porous, irregular-shaped particles, to remove actinides and strontium from the high-level waste salt solutions. Although the process details differ for some of the cesium removal options, in its simplest form, the process combines MST with the high-level waste salt solutions in a reaction vessel, mixing (typically for 24 to 48 hours) to promote the sorption of strontium and actinides with the MST solids, and then filtering to separate the MST solids from the "decontaminated" salt solutions. The MST solids are then washed to remove residual salt solutions and are transferred to the Defense Waste Processing Facility (DWPF) for additional processing and eventual immobilization in borosilicate glass. The MST step is performed in a separate reaction vessel for the ion exchange, solvent extraction, and direct grouting options. For the small-tank TPB option, the MST step is carried out concurrently with the cesium removal step in a single reaction vessel.

To operate successfully, the baseline process must meet the following requirements:

• After treatment with MST, the "decontaminated" salt solutions must meet the strontium and actinide limits shown in Table 3.1 to be acceptable for disposal in the onsite saltstone facility.
• The process must deliver sufficient feed of MST solids to the DWPF to provide for continuous operation of the glass melter. To this end, MST sorption kinetics for strontium and actinides, MST solids filtering, and MST solids washing must be rapid enough to support the required process cycle times.
• The MST solids feed to the DWPF must have a composition that is compatible with the DWPF glass. In particular, the concentration of titanium (Ti), must be less than the limits established for the DWPF, or else the feed will have to be diluted, resulting in the production of additional glass canisters.

TABLE 3.1 Saltstone Waste Acceptance Criteria for Decontaminated Salt Solutions

Radionuclide	Limit (nanocuries per gram of saltstone)
Strontium-90	40
Plutonium-241	200
Neptunium-237	0.03
Total Alpha	20

SOURCE: Jones (2000b).

During its information-gathering sessions, the committee received written information and briefings on all of these issues, several of which are reviewed below.

STRONTIUM AND ACTINIDE REMOVAL

The mechanisms for strontium and actinide removal by MST are not well understood. Presumably, the removal mechanism involves an ion-exchange reaction of the sodium ions in the MST, primarily with cations in higher oxidation states (e.g., strontium, plutonium, neptunium, and uranium), but also, to a lesser extent, with monovalent cesium and potassium cations. The mechanism also may involve the sorption of these cations into the MST structure. Experimental work has demonstrated that under alkaline conditions, MST has a higher removal capacity for strontium and actinides than for other cationic species, particularly cesium and potassium, that are present in the salt solutions. Indeed, MST has been demonstrated empirically to remove strontium and actinides from the SRS high-level waste salt solutions with high efficiency relative to cesium and potassium.

Although MST can be demonstrated to remove strontium and actinides from the tank waste, it is not clear that removal rates are sufficiently high to provide the needed throughput to the DWPF, especially at low MST concentrations. This is particularly true for salt solutions with high ionic strengths (i.e., high Na^+ concentrations), or salt solutions with high plutonium-238 concentrations. In the 1983 in-tank precipitation (ITP) demonstration and the 1995 startup operations (see Chapter 4), MST was used successfully to remove strontium and actinides from salt solutions in Tank 48. However, these solutions apparently contained low concentrations of strontium and actinides and therefore met the saltstone limits (Table 3.1) without MST processing. The MST concentrations used in these operations were 0.6 gram of MST per liter of salt solution in the 1983 demonstration and 1.1 grams of MST per liter of salt solution in the 1995 startup operations. As discussed elsewhere in this chapter, these MST concentrations may be too high to meet DWPF glass limits.

Work on MST kinetics following 1995 startup operations has employed both real and simulated salt solutions using lower concentrations of MST (0.2 to 0.4 gram MST per liter of salt solution) compared to the 1983 demo and 1995 startup. Perhaps the most significant finding from this work is that removal rates differ for the various cationic species of interest. At 0.2 gram of MST per liter of salt solution, removal rates for actinides are lower than for strontium. Indeed, SRS representatives reported to the committee that neptunium removal rates appear to be insufficient to meet the saltstone limits (Table 3.1) after 24 hours of contact. Moreover, plutonium removal rates appear to change significantly after about 10 hours of contact with MST, suggesting that this actinide exists in more than one oxidation state in the salt solutions. Uranium removal is not a major concern, owing to its low concentrations in the salt solutions, although it does consume some of the sorption capacity of the MST and, therefore, can have an effect on strontium and actinide removal rates.

There are several possible ways to increase actinide removal rates to meet the saltstone limits and DWPF throughput requirements. For example, MST concentrations can be increased. SRS representatives reported to the committee that doubling the MST concentration (to 0.4 gram of MST per liter of salt solution), for example, is sufficient to ensure neptunium and plutonium removal for the "average" actinide concentrations encountered in the tank waste. To obtain the required levels of Pu removal, however, it may be necessary to dilute the salt solutions from tanks having high actinide concentrations with salt solutions from tanks having lower actinide concentrations, a process known as blending. This would require an extra processing step, potentially involving the transfer and mixing of waste from several tanks prior to MST processing. Additionally, reaction vessel size can be increased to provide for longer contact times. Experimental work suggests that the required levels of actinide removal can be obtained by increasing the planned reaction vessel size by 25 percent (from the current baseline design of 100,000 gallons [380,000 liters] to 125,000 gallons [470,000 liters]) and diluting the salt solutions from 6.4 molar Na^+ to 5.6 molar Na^+ through the addition of water.

MST SOLIDS REMOVAL

Once MST solids have been added to the salt solutions to sorb strontium and actinides, these solids (along with any sludge solids in the waste) must be separated from the liquids and transferred to the DWPF. The process for removal of MST solids from the tank waste has been described by SRS as having a potentially significant impact on the success of the strontium and actinide removal step, especially for the ion exchange, solvent extraction, and direct grout options. The baseline technology for MST solids removal is crossflow filtration, in which the MST-sludge slurry is streamed tangentially across the face of a microporous filter. The MST and sludge solids are retained in the slurry stream, whereas the decontaminated salt solutions pass through the filter. Tangential flow helps reduce filter clogging.

According to SRS, crossflow filtering provided an acceptable rate of solids removal in both the 1983 ITP demo and the 1995 ITP startup operations. Both of these operations, however, involved the co-filtering of MST, sludge, and TPB solids. SRS characterized the filter performance in the 1983 demonstration as "good" and the filter performance of the 1995 startup operations as "excellent." The improved filter performance in 1995 startup was attributed to enhancements in the filter design and a narrower size distribution of MST solids, which helped prevent filter clogging.

Filtration performance for slurries containing only MST and sludge solids (i.e., without TPB solids) exhibits a 2- to 3-fold decrease compared to MST-TPB-sludge filtering. SRS suggested that a possible reason for this difference was that the larger particle sizes of the TPB solids reduces filter clogging. There are several ways to increase MST and sludge removal rates to achieve needed throughputs should the ion exchange, solvent extraction, or direct grout options be selected. For example, filter size can be increased, flocculents and filter aids [e.g., poly(ethylene oxide), copolymers, or bentonite] can be used to enhance filter performance, or alternate separation technologies (e.g., involving density separations) can be employed. Several of these alternatives are being examined at SRS.

GLASS COMPATIBILITY

MST exhibits good phase stability in the alkaline high-level waste salt solutions at SRS. Consequently, almost all of the MST solids that are added to the processing vessel would be in the waste feed to the DWPF, and very little dissolved MST will end up in the decontaminated salt solutions that are sent to the saltstone facility. The issue of MST compatibility in the DWPF glass focuses on the addition of Ti, which can reduce glass durability and change its processing properties.

The current limit for TiO_2 in feed to the DWPF is 1 weight percent. SRS estimates that at the currently planned MST treatment levels of 0.2 to 0.4 grams of MST per liter of salt solution, the concentration of TiO_2 in the DWPF would range from 1 to 2 weight percent (Dimenna et al., 1999, p. 49). If the crystalline silicotitanate ion exchange option were selected, however, even more TiO_2 would be sent to the DWPF—between 2.5 and 4 weight percent (Dimenna et al., 1999, p. 134). In other words, all of the current processing options appear to exceed current DWPF glass limits for TiO_2 (Hobbs, 2000).

SRS is aware of and has begun to study this problem (Edwards, Harbour, and Workman, 1999a). In briefings to the committee, SRS representatives presented data suggesting that the increased TiO_2 loading would not cause significant changes in glass durability (Edwards, Harbour, and Workman, 1999b). Site personnel acknowledged, however, that a complete variability study would need to be done to ascertain the effects of increased TiO_2 loading on glass durability, homogeneity, liquidus temperature, and viscosity, and that DWPF glass formulation models would need to be updated to reflect this information before any of the processing options could be implemented. There appeared to be some difference of opinion among SRS staff on the cost and time for completing this study.

MST AVAILABILITY

If SRS selects MST for strontium and actinide removal, it would have to procure this material in large quantities from a commercial vendor. Two suppliers have indicated interest in production of MST, and one has made sufficient amounts to validate their process. There is no history of problems resulting in the production of substandard batches. To the committee's knowledge, however, SRS has not yet established detailed material specifications for MST, so the potential for future manufacturing problems remain uncertain.

ALTERNATE PROCESSES

The committee's interim report noted that SRS was not considering alternative materials to MST for removal of strontium and actinides (National

Research Council, 1999b, p. 7). Although SRS personnel reported that they were confident that MST performance would be sufficient to meet the processing needs for the small tank precipitation process, the committee learned at its November 1999 meeting that alternate processes were being reviewed in case MST fails to meet expectations. Among alternate processes to MST that are being considered are the following: sorption by sodium nonatitanate ($Na_4Ti_nO_{2n+1}$ or $Na_4Ti_nO_{2n+2}$), or SNT; ferric hydroxide flocculation; permanganate reduction; and sodium uranate formation. Some testing is planned to determine the ability of these alternate processes to achieve required decontamination levels. It also was suggested that it might be possible to coat the crystalline silicotitanate (CST) particles that would be used in the ion exchange process with an actinide-absorbing agent. However, no further details of this possibility were provided to this committee.

Among the reasons for choosing MST is its capacity for actinides, which is adequate at reasonable loadings. Importantly, the loading at full capacity is insufficient to have a criticality incident. The Allied Signal (now Honeywell) SNT is an appealing alternative from the standpoint of reliable sources because it is produced in large amounts for a worldwide market. If SNT is an acceptable alternative, a range of possibilities can be considered—for example, SNT could be used in an ion exchange mode, either alone or in some combination with crystalline silicotitanate (see Chapter 5), which would mean that filtration might be limited to sludge alone. Filtration is a consideration since, as discussed previously, MST has a 2-3 fold increase in filtration efficiency if the filtration is carried out in the presence of the TPB precipitate.

R&D ACTIVITIES TO RESOLVE UNCERTAINTIES

In its briefings to the committee (Westinghouse Savannah River Company, 1999a, 1999c) and in written responses to subsequent committee questions (Jones, 2000a), SRS has provided the committee with an outline of its planned R&D activities to resolve the remaining uncertainties with the strontium and actinide removal step. The R&D plans were being completed as this report went to review. The committee did not have the opportunity to obtain or perform a detailed evaluation of these plans.

The planned R&D work falls into the following broad categories:

• MST filtration studies to understand the role of TPB in enhancing filtration of TPB-MST-sludge slurries, bench-scale screening to identify flocculents that can be used to increase filtering rates, and larger-scale filtering tests of these flocculents using real waste.

• Alternate methods to remove MST solids from the decontaminated salt solutions involving the testing of a small number of removal technologies identified in a recent Tank Focus Area report (U.S. Department of Energy, 1999).

- Studies to obtain a more detailed understanding of removal rates of strontium and actinides using MST, and investigation of the ways to increase these rates by varying MST concentrations and processing conditions.
- Evaluation of the efficacy of other materials (sodium nonatitanate, ferric hydroxide, permanganate, sodium diuranate) for removing sodium and actinides from high-level waste salt solutions.
- Evaluation of the effects of plutonium oxidation states on MST sorption rates.
- Determination of concentrations of soluble actinides in the salt solutions through direct measurements of tank samples to aid design of MST processing conditions.

ANALYSIS

In its interim report (National Research Council, 1999b) the committee reached the following conclusion with respect to the strontium and actinide removal step: (a) there appear to be some remaining technical questions that will need to be resolved before this process can be successfully implemented at SRS; and (b) Westinghouse Savannah River Company appears to be pursuing these questions vigorously. The committee also concluded that the information it received subsequent to the completion of its interim report does not justify a change in its position; that is, the committee continues to believe that the MST process holds sufficient promise to justify its continued development. Nevertheless, in the view of the committee, two major issues remain to be resolved before SRS can successfully implement this process:

1) SRS must demonstrate that strontium and actinide removal can be accomplished within saltstone limits and at the throughput rates required by the DWPF; and
2) SRS must demonstrate that the MST concentrations used to remove the strontium and actinides will not exceed compatibility limits for DWPF glass.

That SRS now appears to be considering alternatives to the MST process is perhaps a sign that there is some uncertainty concerning whether MST can be shown to work. The committee concurs with this concern. The consolidation of alternatives represents prudent engineering practice in the face of technical uncertainties.

The committee believes that further R&D is needed to confirm that the MST process can operate successfully within the technical requirements. The plans for R&D work that were presented to the committee in outline form seem appropriate for resolving many of the technical uncertainties with this processing option. As noted above, however, the details of these plans were still being developed as the committee finalized this report; consequently, the committee was unable to evaluate the R&D plans in detail. The committee cannot provide an endorsement of the R&D in the absence of sufficient information that allows evaluation of the details of these plans.

RECOMMENDATIONS

In light of the foregoing discussion, the committee offers the following recommendations to promote the resolution of the technical uncertainties with respect to the MST processing step.

1) SRS should resolve the technical uncertainties with the MST processing step as soon as possible by implementing an R&D program focused on the technical uncertainties noted above, and others that are consistent with a systems engineering approach to salt processing. This R&D should provide a demonstration that MST functions satisfactorily with each of the cesium separation process options.

2) Simultaneously, a well-focused R&D program should be conducted to examine alternatives to MST. This program should have a level of effort commensurate with the risk of process failure and should continue until MST processing can be demonstrated to meet the saltstone, DWPF throughput, and DWPF glass requirements.

3) As part of its efforts to resolve technical uncertainties, SRS should establish requirements for and reliable sources for the manufacture of MST.

4

Tetraphenylborate: In-Tank Precipitation and Small-Tank Precipitation Options

The focus of this chapter is on the tetraphenylborate (TPB) precipitation process, which was developed at Savannah River Site (SRS) in the late 1970s and early 1980s to remove cesium from high-level waste (HLW) supernates. The original design for this process involved the use of an existing underground HLW waste tank at the site; consequently, the process was referred to as *in-tank precipitation*, or ITP. The in-tank precipitation process was abandoned in 1998 because of technical difficulties, and a hybrid process, referred to as *small-tank TPB precipitation*, was developed as a potential alternative. This chapter provides a review of these processes and identifies remaining scientific and technical difficulties. Several recommendations on the implementation of the small-tank precipitation process are provided at the end of the chapter.

TETRAPHENYLBORATE PRECIPITATION PROCESS

The TPB precipitation process removes cesium from the supernate by precipitation with sodium tetraphenylborate, $Na[B(C_6H_5)_4]$, through the following reaction:

$$Cs^+_{(aq)} + Na[B(C_6H_5)_4]_{(aq)} \rightleftharpoons Cs[B(C_6H_5)_4]_{(solid)} + Na^+$$

where the double arrows indicate that the reaction is reversible.

Sodium tetraphenylborate (NaTPB) is a reagent with well-known properties. The low solubility of CsTPB (the solubility product, or K_{sp}, at 25 °C is 7.84×10^{-10}) potentially provides decontamination factors as high as 10^5 to 10^6, and the CsTPB precipitate is typically in a form that is easily filtered.

As originally designed for the ITP process, NaTPB and monosodium titanate (MST) (see Chapter 3) were to be added to the HLW supernate to precipitate cesium, strontium, and actinides. The precipitate was to be removed from the tank by filtration and was then to be treated to remove greater than 90 percent of the organic material (i.e., the phenyl $[C_6H_5]$ groups bound to boron) through a precipitate hydrolysis process using formic acid in the presence of a copper catalyst (Ferrara, Bibler, and Ha, 1992). The products of this reaction are benzene, which can be removed by evaporation and subsequent incineration, and an aqueous solution containing Cs, K, and $B(OH)_3$ ions. This aqueous solution was to be fed to the Defense Waste Processing Facility (DWPF) to be incorporated into glass, and the decontaminated supernate was to be incorporated into grout at the SRS Saltstone Facility. A schematic flow sheet for this process is shown in Figure 4.1.

In-Tank Precipitation Process

In the original design, the TPB precipitation process was to be performed in an existing HLW tank at SRS, and a large-scale test in an actual HLW tank was conducted in 1983 to demonstrate proof-of-principle. The test was conducted in Tank 48, a 1.3-million gallon (5-million liter) underground storage tank in the H-Tank Farm (see Walker et al., 1996). MST and TPB were added to the tank waste, resulting in the generation of 20,000 gallons (76,000 liters) of precipitated slurry containing cesium and other metals. During the wash phase of the test, 183,000 gallons (693,000 liters) of water were added to the tank while the slurry pumps were operating. Benzene generation was noted, and benzene levels in the tank exceeded the maximum instrument readings for a period of 6 hours. The SRS review of the experiment concluded that the test was a success, but recommendations were made that the causes for high benzene release rates be investigated.

Additional analyses on the cause(s) of the benzene generation resulted in an *incorrect* conclusion in 1983: namely, that benzene generation was due primarily to radiolysis. Additional testing at the University of Florida in the mid-1980s under conditions different from those in Tank 48 provided values for radiolytic production for free and trapped benzene (the latter refers to benzene that is physically held within the waste). In 1987 and again in 1994, Savannah River Technology Center (SRTC) conducted tests under conditions similar to those in Tank 48, but failed to duplicate the phenomenon of trapped benzene (Defense Nuclear Facilities Safety Board, 1997, Section 2.3.2). Additional work at the Georgia Institute of Technology confirmed the SRTC findings that the 1983 conclusion on the benzene generation mechanism was incorrect. Nevertheless, the committee understands that over the next 12 years, no comprehensive studies were initiated to identify the mechanism(s) of benzene generation and release or to examine its potential effects on ITP processing requirements.

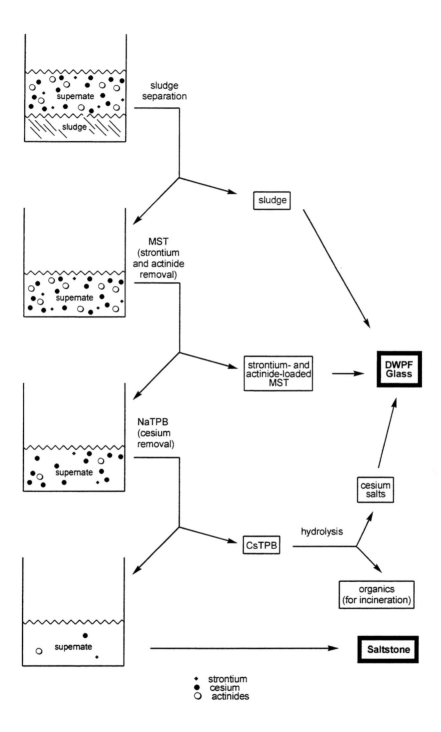

FIGURE 4.1 Schematic flow sheet for radionuclide removal from high-level tank waste at Savannah River using small tank TBP precipitation.

From the late 1980s to the early 1990s, chemists at SRS were engaged in NaTPB stability tests, since slow decomposition of vendor-supplied NaTPB had been observed. Although NaTPB is typically supplied as a 0.5 molar solution stabilized by 0.1 molar NaOH, it was established that, even in alkaline solution, copper ions catalyze the decomposition of TPB to benzene, phenol, and boric acid (Barnes, 1990, 1992; Crawford et al., 1999). A soluble copper impurity was identified along with other metals, for example, Pd and Ni, in the vendor-supplied NaTPB. Another research program addressed the problems of waste foaming during the ITP processing, which was of concern because of the potential for clogging transfer pipes and inhibiting phase separation.

In 1995, a large-scale production operation was conducted primarily to assess potential benzene vapor phase mixing, temperature, and oxygen concentration in Tank 48. Approximately 1.4×10^5 liters of solution containing 27,500 kilograms of NaTPB were added to Tank 48 over the period from September 2 to September 29, 1995. These added solutions contained an estimated 14.6 kg of dissolved benzene. Beginning on October 9, three slurry pump tests and a filtration test were conducted. During the third slurry pump test on November 9, a maximum temperature of 52 °C was reached in the tank. On December 1, all four slurry pumps were operated to mix the tank contents, and an alarmingly high concentration of flammable benzene vapor that exceeded ten percent of its lower flammability limit accumulated in the tank headspace. The slurry pumps were shut down to slow benzene release from the tank waste.

Over the next few months, slurry pumps were activated occasionally, and high benzene vapor concentrations were repeatedly noted. An estimated 8,500 kg of benzene was generated during the period November 5, 1995 to April 22, 1996 and was eventually removed from Tank 48. Subsequent analysis indicated that greater than 95 percent of the NaTPB decomposed over the period from November 14 to December 28. Depending on which time is assumed for completion of this decomposition, the TPB decomposition rate was estimated as either 25,000 μg/liter·hr (upper bound) or 12,500 μg/liter·hr (lower bound), both far in excess of that expected from radiolysis (Walker et al., 1996).

Extensive analyses of the contents of Tank 48 led to a satisfactory organic material balance,[1] indicating that most of the TPB decomposed to benzene with lesser amounts to phenylboronic acid, phenol, biphenyl, and much less to terphenyl and diphenyl mercury. Time profiles for concentrations of hydroxide, phenylboronic acid, phenol, potassium, cesium, and boron in Tank 48 were developed for the period of the 1995–1996 benzene evolution. After the DNFSB report, a research program also was initiated by SRS to examine the chemistry responsible for the rapid TPB decomposition

---

[1] That is, the total amount of organic material (i.e., phenyl groups) added to the tank could be accounted for by reaction products (e.g., benzene, phenol, biphenyl) and residual TPB after the excursion.

in the tank. Since copper-catalyzed decomposition of TPB had already been developed for downstream processing of the TPB sludge (the precipitate hydrolysis process mentioned previously in this chapter), and since copper was identified as a catalyst resulting in slow decomposition of the vendor-supplied NaTPB, copper was an initial suspect as a catalyst responsible for decomposition.

Using Tank 51H sludge and/or Tank 48H filtrate, nine TPB decomposition tests revealed decomposition rates as high as 2,214 µg/liter·hr, and more commonly 30 to 773 µg/liter·hr at 40 °C. At 70 °C, in the presence of oxygen, TPB decomposition displayed a lag time of several days. In the absence of oxygen, decomposition was initiated immediately. However, in none of these tests, including those in which suspected metal ion catalysts were added, were rates observed that were comparable to those seen in the 1995 Tank 48 excursion. Moreover, only a small percentage of the NaTPB was decomposed in the test runs, leading researchers to suspect that most of the benzene evolution in these tests resulted from impurities in the vendor-supplied NaTPB. In no case was complete NaTPB decomposition observed, again in contrast to the excursion observed in the 1995 production operations.

An outside panel of experts, the Process Chemistry and Mechanisms Panel, was established in January, 1996 to guide an experimental program to determine the mechanism(s) of decomposition of NaTPB and benzene release involved in the 1995 excursion. The panel was composed of Robert Hanrahan (University of Florida), Bruce King (University of Georgia), Edward Lahoda (Westinghouse Science and Technology Center), George Parshall (DuPont Central Research, retired), and Robert Smiley (DuPont, retired), with occasional participation of SRS consultants Russell Drago (University of Florida) and Preetinder Virk (Massachusetts Institute of Technology). The experimental program work was conducted by approximately ten Westinghouse Savannah River Company (WSRC) scientists and engineers, although only six had a primary assignment to this project. Some additional experiments were carried out by other researchers at Georgia Institute of Technology, DuPont, and Pacific Northwest National Laboratory. The experimental work was periodically reviewed and research plans were modified based on discussions with members of the Process Chemistry and Mechanisms Panel, which itself produced 16 reports during its two years of existence (Process Chemistry and Mechanisms Panel, 1996a-k, 1997a-d, 1998).

Most of the tests were conducted with non-radioactive simulants to screen various metal ions for catalytic activity toward TPB decomposition. Based on the analysis of the contents of Tank 48, a simulant slurry recipe was developed for non-radioactive testing. Copper-catalyzed decomposition was suspected during the initial stages of the program (see Crawford et al., 1999), and various tests were conducted using simulated waste and accelerated conditions, such as higher temperature and greater concentrations of suspected metal-ion catalysts. Many essential features of the 1995 event could not be duplicated. Notably, the observed lag time of more than two months preceding rapid TPB decomposition was not duplicated in the tests.

More importantly, the copper catalyst activities were far too low (by at least two orders of magnitude) to account for the rates of benzene released in the 1995 excursion, even at the 52 °C maximum temperature experienced in the 1995 production operations. The tests with simulants revealed benzene release rates that were even lower (by approximately one order of magnitude) than the rates measured with radioactive Tank 51H sludge and/or Tank 48H filtrates.

On August 14, 1996, the Defense Nuclear Facilities Safety Board (DNFSB) recommended that the Department of Energy not proceed with large-scale process testing at the ITP Facility until the mechanisms of benzene generation, retention, and release were better understood and adequate safety measures had been developed to mitigate benzene deflagrations (Defense Nuclear Facilities Safety Board, 1997). Planned operations were put on hold while research on catalytic mechanisms continued.

In the fall of 1996, the Process Chemistry and Mechanisms Panel began to question the possible role of noble metal (e.g., palladium) catalysis. Early in 1997, a significant catalytic influence from the combined presence of noble metals, select organic compounds, and a tetraphenylborate precipitate was noted. As compared with copper catalysis, greater TPB decomposition rates were obtained in simulated tests with palladium additives, but the greatest rates required much higher levels of palladium than have been found in real waste and also required addition of diphenyl mercury, TPB decomposition intermediates, and benzene. Additional features were cited in the Panel's reports for this catalyst system (Process Chemistry and Mechanisms Panel, 1998):

- activation of the catalytically active species is slowed in the presence of oxygen;
- temperature, radiation, and copper synergism influence catalyst activation; and
- tetraphenylborate solids provide a support platform for the palladium catalyst system.

A working mechanism was developed that invokes initial TPB reduction of soluble palladium ions to elemental palladium [Pd^{2+} to $Pd(0)$], interaction of $Pd(0)$ with TPB to form a phenylpalladium intermediate that is subsequently protonated by water to benzene and triphenylboron, which regenerates $Pd(0)$. A similar cycle catalyzes hydrolysis of triphenylboron. Copper is postulated to catalyze the subsequent decomposition of tetra-, tri-, and diphenylborons and phenylboronic acid.

At its final (16th) meeting in March, 1998, Process Chemistry and Mechanisms Panel (1998) provided the following summary comments:

- the key catalyst is palladium metal deposited on a solid support (TPB solids, sodium titanate, or sludge solids);
- reaction initiation is affected by several parameters which control redox state (oxygen, temperature, intermediates, mercury, TPB, and copper);

- the TPB decomposition mechanism involves two energies of activation;
- the decomposition mechanism results from soluble TPB interacting with the catalyst; and
- the lower TPB solubility in the presence of potassium is most likely due to the formation of a mixed crystalline form.

The Panel was supportive of additional testing to enhance technical understanding of the decomposition mechanism, especially in the areas of temperature effects, induction period, oxygen, and the use of inhibitors.

Also, in early 1998, SRS concluded that both safety and production requirements could not be met using the ITP process, and a decision was made to suspend operations and search for alternatives. At the time of suspension, SRS had spent $489 million to develop and implement the ITP process (U.S. General Accounting Office, 1999, p. 4).

Small-Tank TPB Precipitation

The small-tank TBP precipitation process shares many of the features of the ITP process, except that it is carried out in smaller, purpose-built tanks to provide greater control over precipitation and benzene formation. The process allows for closer temperature control and faster cycling times to reduce the generation of benzene and improved agitation of the liquid to facilitate benzene removal. The process is also designed with secondary containment and positive pressure control so that the processing vessels could be blanketed with nitrogen to reduce explosion hazards and facilitate benzene removal.

Design of the small-tank process has focused primarily on engineering issues: minimization of waste foaming, optimal mixing conditions to minimize the amount of NaTPB needed, measurements of precipitation and dissolution rates, and establishment of the conditions that optimize the recovery of excess NaTPB during washing. Work on many of these issues was underway during this committee study, and the committee received briefings on some of this work during its information-gathering meetings.

Foaming involves the formation of microscopic gas bubbles in the waste during agitation of the waste-TPB slurry. Excessive foaming can disrupt waste processing operations; the foam can inhibit phase separation and disrupt flow through waste-transfer lines. The waste foaming problem was first discovered in laboratory experiments utilizing real samples of tank waste. Current research efforts at the SRTC and Oak Ridge National Laboratory are focused on identifying the causes of foaming, identifying foam-control reagents, testing the radiolytic stability of these reagents, and determining the potential impacts of these reagents on downstream process operations, especially the DWPF.

ANALYSIS

In view of the very low solubility of CsTPB, it was logical for SRS to develop processes based on TPB precipitation for removing cesium from the HLW supernate. In laboratory applications, large decontamination factors have been demonstrated, and the precipitate is typically in a form that is easily filtered. Further, NaTPB is relatively inexpensive, and subsequent acidification of the CsTPB precipitate (the precipitate hydrolysis process discussed elsewhere in this report) allows for controlled decomposition to easily separated benzene that can be destroyed in an existing incinerator, and an aqueous stream containing boric acid and cesium and potassium salts suitable for vitrification at the DWPF.

The implementation of this process at SRS, however, was flawed for a number of reasons. Benzene production at higher-than-measurable levels during the 1983 test in Tank 48 was not correctly analyzed, and incorrect conclusions concerning the generation and release of benzene were reached. As noted in a previous section, work at SRTC and Georgia Institute of Technology called into question the correctness of the postulated benzene generation mechanism as early as 1987. The 1995 production operations in Tank 48, which were conducted to evaluate whether the benzene vapor could be adequately mixed for safe operation and to test filtration procedures, produced an unexpected, rapid TPB decomposition. The causes of this rapid TPB decomposition have not yet been identified. The discovery process is complicated by the fact that the contents of Tank 48 at the time the excursion were highly complex. They included not only the supernate, but also sludge together with much of the tetraphenylborate precipitate and its various decomposition products from the 1983 test.

As was noted earlier in this chapter, early attempts were made to identify the catalyst(s) responsible for the TPB decomposition in Tank 48, and copper catalysis was considered a likely candidate. However, at least two important features could not be reproduced in tests with tank filtrate and/or sludges, even when elevated concentrations of suspected catalytically active ions were added: (1) the rapid rates of benzene formation, (2) complete decomposition of TPB observed in Tank 48. Tests with non-radioactive simulants afforded even lower rates of TPB decomposition, even at higher copper concentrations than those present in Tank 48 (Crawford et al., 1999). Palladium was later found to produce higher TPB decomposition rates, but only in the presence of other phenylated additives, such as diphenyl mercury, TPB decomposition intermediates, and benzene. It is not clear whether these higher rates are sufficient to account for the observed 1995 excursion. The palladium-catalyzed decomposition might be explained by synergism of Pd(0) with the additives, as well as an additional catalytic cycle using copper.

By contemporary standards of mechanistic understanding in organometallic chemistry, the current proposed mechanistic scheme for TPB decomposition is rather crude, and some of the key proposed steps do not appear to have precedent. A variety of transition metal ions have been shown to stoichiometrically oxidize TPB to triphenylboron and biphenyl: Ce(IV), Ir(IV), Fe(III), and Cu(II) (Eisch and Wicsek, 1974; Turner and Elving,

1965; Geske, 1959, 1962; Abley and Halpern, 1971; Strauss, 1993), or to catalyze TPB hydrolysis (Flaschka and Barnard, 1960). Truly catalytic hydrolytic decomposition of TPB in alkaline solution appears to be confined to copper-based systems (Barnes, 1990, 1992; Crawford et al., 1999) and a palladium/phenylated boron/diphenyl mercury/benzene system discovered by SRS scientists (see above).

The first step in the proposed palladium catalysis system involves what appears to be transfer of a phenyl group from boron to Pd(0). Such transmetallation chemistry is well established for a wide variety of metal ions: Fe(II), Fe(III), Ni(II), Zr(IV), Rh(I), Rh(III), Ru(II), W(IV), Hg(II), Pt(II) (Bianchini et al., 1989; Bonnessen et al., 1989; Legzdins and Martin, 1983; Reed et al., 1979; Sacconi, Dapporto, and Stoppioni, 1976; Haines, and duPreez, 1971; Clark, and Dixon, 1969), and Pd(II) (Cho and Uemura, 1994; Cho, Ohe, and Uemura, 1995; Crociani et al., 1990, 1991; Moreno-Mañas, Pérez, and Pleixats, 1996), but not, to our knowledge, for Pd(0). Also unclear are the mechanisms by which diphenyl mercury and the decomposition intermediates accelerate the palladium-catalyzed decomposition sequence.

The lack of understanding of the details of the palladium catalytic cycle, or for that matter, whether a palladium system is responsible for the TPB decomposition in Tank 48, remain matters of concern, since the possibility exists that another, perhaps even more rapid TPB decomposition scenario could be repeated in a future processing operation. The apparent variability of tank waste composition at SRS (see Chapter 8) raises additional concerns, especially given that the reasons for rapid TPB decomposition in Tank 48 remain unexplained. It seems unlikely that a single mechanism for TPB decomposition will explain the 1995 excursion and, at the same time, will foreshadow all possible decomposition scenarios with waste from the other tanks.

Apart from the lack of mechanistic understanding, the failure to duplicate the high rates of TPB decomposition in any of the tests, with real tank waste or simulants, illustrates the current lack of understanding of the tank waste chemical system. The estimated average rates for TPB decomposition in Tank 48 are based on an assumed steady decomposition over a period of approximately six weeks. However, a very high rate, far in excess of the assumed 25,000 µg/liter·hr (upper bound), but for a much shorter period, cannot be ruled out. Although 25,000 µg/liter·hr rates have not been achieved in any of the tests, the actual rate for the TPB decomposition in Tank 48 is unknown. Moreover, the observed months-long lag between TPB addition and maximum benzene production in Tank 48 in the 1995 production test also has not yet been adequately explained or duplicated in any subsequent tests.

There appears to have been over-reliance on tests with simulants in the past research programs addressing TPB decomposition mechanisms. Early tests with Tank 51 sludge and Tank 48 filtrate produced significantly greater TPB decomposition rates, as compared with tests with simulants. Given the complex compositions of the tank sludge, filtrate and solutions, it is not surprising that the principal components responsible for faster TPB decomposition have probably not been identified.

The rapid decomposition of NaTPB that occurred in the 1995 processing operations has several consequences that weigh against its use for removal of cesium by the ITP process as originally planned. One major difficulty is the generation of large amounts of benzene, which could present safety problems if not properly handled. Although safety issues could presumably be resolved by standard industrial processing controls, the quantity of benzene generated could pose regulatory problems. Another problem is the reduction in decontamination factor. As the soluble TPB decomposes, and hence the concentration of $[B(C_6H_5)_4]^-_{(aq)}$ decreases, the equilibrium would shift by redissolution of precipitated cesium tetraphenylborate as indicated by the equation shown at the beginning of this chapter. Any $Cs[B(C_6H_5)_4]$ that redissolved would lower the decontamination factor, and this could be counteracted only by addition of more NaTPB. Such further addition would be inefficient and would merely revert to an earlier stage of the process, with all of the potential problems of decomposition unchanged. In other words, the 1995 excursion might not represent an extreme rate at all, and even higher rates of TPB decomposition might be possible, especially as tank heels (hard-packed waste at the bottom of the tanks) of varying composition are processed.

Development of a small tank TPB alternative appears to be an attempt to "engineer around" the problems observed with the ITP process. Although the final design is not yet complete, SRS believes that improved mixing and temperature control would permit much shorter residence times for the CsTPB precipitate (estimated currently as several days), as compared with the 200-day processing and up to two-year storage time for the batch-type ITP process. These significantly reduced processing times, of course, similarly reduce the possibility that a TPB catalytic decomposition catalyst system might evolve. Moreover, some TPB decomposition is assumed in the current design of the small tank process: slower feed rates and recycle of off-spec material to reprecipitate any soluble cesium are available contingencies. Nevertheless, the lack of a mechanistic understanding of the TPB decomposition process or empirical bounds on decomposition rates present significant hurdles to the successful implementation of TPB precipitation in the proposed small-tank process.

FINDINGS AND CONCLUSIONS

Based on the foregoing analysis, the committee identified five findings and conclusions with respect to cesium removal using the small tank TPB process:

1) Given the very low solubility of CsTPB, it is understandable that SRS has explored TPB precipitation as a strategy for removing cesium from the HLW supernates at the site. Large decontamination factors are possible in laboratory applications, and the precipitate typically exists in a form that is easily filtered. Further, NaTPB, as is available at this time, is relatively inexpensive and pure, and subsequent acidification of the CsTPB precipitate al-

lows for controlled decomposition to benzene, which can be incinerated, and an aqueous stream containing boric acid and cesium and potassium salts suitable for vitrification in the DWPF.

2) Many of the scientific, technical, and regulatory issues associated with TPB precipitation have been identified. The regulatory issues center primarily on maintaining benzene release at or below regulatory limits and below its flammability limit. The scientific and technical issues include the following:

- MST adsorption and ion exchange kinetics (see Chapter 3).
- Decontamination factors for cesium under small-batch semicontinuous or continuous process operation.
- Acceptance criteria for the ultimate glass waste form (see Chapter 8).
- Precipitate washing and recycle of NaTPB.
- Excessive foaming of stirred radioactive waste slurries.
- Cycle times and products associated with the decomposition of CsTPB during precipitate hydrolysis processing.

3) The formation of foam in stirred laboratory experiments with NaTPB and HLW and its implications with respect to clogging of transfer pipes *and* poor phase separation may be a major impediment to the use of this technology in the small tank TBP process.

4) Despite conclusions to the contrary (Process Chemistry and Mechanisms Panel, 1998), the causes of rapid TPB decomposition in the 1995 production operations in Tank 48 remain uncertain. SRS has not achieved a level of understanding sufficient to prevent TPB decomposition, choosing instead to "engineer around" the problem. Although the completed research on the copper-catalyzed TPB decomposition scheme (Crawford et al., 1999) is a valuable contribution, the studies have not really addressed the key issues relevant to ITP or the proposed small tank variant. Copper is one of many potential catalysts, and the tank waste is sufficiently complex and heterogeneous that a mechanistic understanding of catalysis is probably not possible given the time and resources available to this project.

Although the design of the small tank TPB process appears to considerably reduce the likelihood of an event analogous to the 1995 excursion, it may not be possible to entirely prevent future rapid TPB decomposition. Since the contents of individual tanks are not homogeneous and contents can vary substantially from tank to tank, some of the waste could have much higher concentrations of the catalyst(s) for TPB decomposition than were present in Tank 48.

RECOMMENDATIONS

Based on these findings and conclusions, the committee offers the following four recommendations. They are directed primarily to the decision makers and research managers responsible for the cesium separation program:

1) If small-tank TPB precipitation remains as a contending process for removing cesium from HLW supernates at SRS, considerable effort should be made to (i) identify TPB catalytic decomposition mechanisms, (ii) establish probable bounding rates for TPB catalytic decomposition, and (iii) address the other scientific and technical issues listed in finding 2 above.

2) As part of its efforts to bound catalytic decomposition rates, SRS should develop robust testing protocols to process moderately sized samples of real waste from each of the tanks using MST and TPB. These protocols are needed to provide information that can be used to develop a predictive capability for process performance for TPB decomposition rates, temperature excursions, foaming, and filterability.

3) Tests on moderately sized samples of real waste should be implemented as soon as possible under this protocol to help assess the viability of the small tank TPB precipitation option. By using real waste from different tanks with different compositions, this testing will allow process performance to be systematically established under conditions that bracket, with safety margins, the acceptable conditions of planned processing operations.

4) If the small-tank TPB processing option is selected for implementation at SRS, samples of each of the waste batches to be processed should be subjected to the testing protocols described above. This will help ensure that unknown synergistic effects of waste blending from different tanks are better understood prior to full-scale processing.

5

Crystalline Silicotitanate Ion Exchange

Ion exchange has been in commercial use for over 100 years to remove ionic species from aqueous solutions (e.g., deionization of water). As illustrated in Figure 5.1, a typical process for exchange of cations employs insoluble polymeric beads (inorganic materials, such as zeolites, are also widely used) that have a number of exchangeable sites (shown in the Figure 5.1 as either OH or ONa). Ion exchange processes typically involve equilibrium reactions, and the separated ions are usually *eluted* from the ion exchange material using a dilute acid (e.g., elution of sodium ions in the reverse reaction for Figure 5.1) or a salt solution (e.g., elution of cesium ions in the reverse reaction of Figure 5.1). The elution sequence allows the ion exchange material to be reused in multiple cycles.

Although the underlying technology is well established, ion exchange for cesium removal from high-level waste at the Savannah River Site (SRS) and other U.S. Department of Energy (DOE) sites poses many challenges. The ion exchange material must withstand both high alkalinity and high radiation fields, while at the same time exhibiting selectivity for cesium in the presence of much greater concentrations of chemically similar ions such as sodium and potassium.

A promising ion exchange material, crystalline silicotitanate (CST), was investigated by workers at Sandia National Laboratory and Texas A&M University (Sherman, 1999). CST is an outgrowth of earlier work at Sandia on amorphous hydrous titanium oxide (HTO) in the 1960s and 1970s (Walker, Taylor, and Lee, 1999). An ion exchange medium based on CST, known as TAM-5, was developed in the early 1990s at Sandia and Texas A&M under the auspices of the DOE Environment Management—Office of Science and Technology. Subsequent product development and commercial manufacture was carried out by UOP of Des Plaines, Illinois, under a cooperative research and development agreement with Sandia (Sherman, 1999). The commercial material is sold under the name IONSIV®. The product in

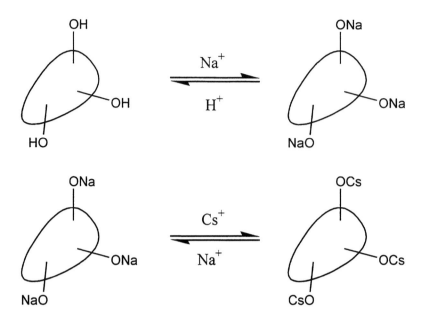

FIGURE 5.1 Schematic depiction of the ion-exchange process.

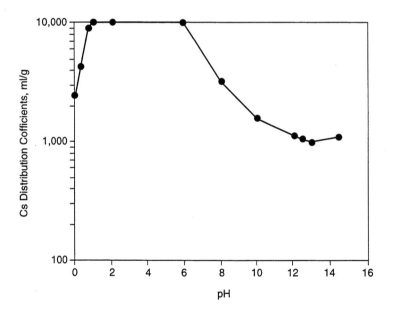

FIGURE 5.2 Selectivity for Cs in CST as a function of pH. A large distribution coefficient indicates high Cs selectivity. SOURCE: Anthony et al. (1994); Zheng, Gu, and Anthony (1995).

current use is designated as IONSIV® IE-911. An earlier product, which was a powdered form of the material, was called IONSIV® IE-910.

Silicotitanates have received considerable attention as ion exchange materials for nuclear waste applications, in particular because they exhibit very high selectivity for cesium ions (Cs^+) and also show specificity for strontium. CST is unusual in that it exhibits high selectivity for cesium ions (Gu et al., 1997) in salt solutions (Anthony et al., 1994) and for cesium over a wide pH range from acidic to basic solution as shown in Figure 5.2 (Zheng, Gu, and Anthony, 1995; Anthony et al., 1994). It also exhibits high stability to radiation (Zheng et al., 1996). Though maximum selectivity is reported at pH ≤6, the reported distribution coefficient (K_d) for cesium under alkaline conditions is still high in comparison to other ion exchange materials. According to Anthony et al. (1994), the decrease in Cs^+ selectivity at high pH is not desirable in applying TAM-5 to defense waste in its current form.

CST is also unusual in that cesium is not easily removed from the material [i.e., the reaction rate for the reverse process (Figure 5.1) is low], and the process is described as *nonelutable* ion exchange. Consequently, it is not practical to recycle the CST, and the loaded CST would need to be incorporated into the high-level waste stream. This, in turn, raises questions of the effects of higher concentrations of titanium on the stability of borosilicate glass.

SRS first evaluated the use of CST for removal of cesium from SRS high-level waste (HLW) using simulants in 1995 and found it to be "very effective for cesium removal" (McCabe, 1995). In 1997, small-scale tests were performed with actual SRS waste (McCabe, 1997). Also, in 1997, SRS collaborated with Oak Ridge National Laboratory (ORNL) to successfully demonstrate a glass wasteform formulation that could incorporate the higher titanium loadings required by the use of nonelutable CST (Andrews and Workman, 1997). SRS has been performing research and development (R&D) on CST as the back-up for small tank precipitation for the last two years.

The proposed CST ion exchange process for treatment of SRS HLW is summarized in Figure 5.3. The process involves the following steps, beginning with separation of the supernate from any sludge in the tank. A slurry of monosodium titanate (MST) would then be added to the waste to sorb strontium, plutonium, and other actinides, and the resulting slurry would be filtered to remove insoluble MST and entrained sludge from the prior step. The insoluble solids would then be washed and transferred to Defense Waste Processing Facility (DWPF) for incorporation into borosilicate glass. The clarified salt solution from filtration would be passed through a series of columns packed with CST to remove cesium. The cesium-loaded CST would be transferred to DWPF for incorporation into glass, while the decontaminated salt solution would be treated as a low level waste and disposed of as saltstone (Figure 5.3).

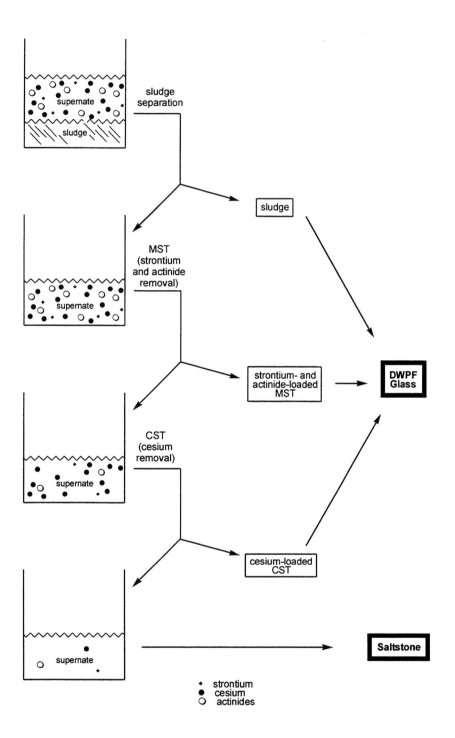

FIGURE 5.3 Schematic flow sheet for radionuclide removal from high-level tank waste at Savannah River using CST ion exchange.

Outside the SRS program, CST in the form of IE-911 has been incorporated into the Cesium Removal Demonstration Project at ORNL. Approximately 1.2×10^5 liters of supernate, obtained from the Melton Valley storage tanks, were processed during the demonstration. Some 265 liters of sorbant were successfully used to remove 1.1×10^3 Ci of ^{137}Cs from the supernate (Walker, Taylor, and Lee, 1999; Lee et al., 1997). The task at SRS is considerably more challenging, with a total free supernate and saltcake volume of 1.2×10^8 liters containing 8.8×10^7 Ci. In other words, the volume of waste to be treated is larger by a factor of 1,000, and the amount of cesium is higher by nearly five orders of magnitude.

In contrast, the Hanford Site used an elutable ion exchange resin to make capsules of cesium. In the early 1980s, the Dupont Chemical Company (then the management and operating contractor at SRS for DOE) conducted a cost analysis of building new facilities for both a vitrification and an ion exchange process. The analysis showed that an ion exchange facility, based on the Hanford process, would cost about the same as the vitrification facility—almost a billion dollars each. By contrast, the tetraphenylborate (TPB) process for cesium removal appeared to offer a way to avoid the cost of a new ion exchange facility by separating the cesium from the waste in the already-existing tanks (U.S. General Accounting Office, 1999). The issue of catalytic decomposition of TPB was not an issue at that time.

PHYSICAL, CHEMICAL, AND MINERALOGICAL CHARACTERISTICS OF CST

Crystalline siliotitanate ion exchange materials have been synthesized in a nominal three-component system Na_2O-TiO_2-SiO_2 (Balmer et al., 1997). A fourth, proprietary dopant component (Savannah River Site, 1999) is used in the commercially available material, and a five-component system is produced by cesium exchange. Partial data in the three-component system, Na_2O-TiO_2-SiO_2, have been reported (Levin, Robbins, and McMurdie, 1979, Figure 531) but there is little information available for the four- and five-component systems. Relatively few details are available on the synthesis of CST. Its preparation by a hydrothermal process has been described (Anthony et al., 1994), but reaction conditions were not reported. A more detailed preparation was reported via a sol-gel route where the precursor materials were mixed for 15 hours and dried in air at room temperature. The amorphous, homogeneous precursor was subsequently heated in air at 800 °C for at least one hour to form crystalline CST (Balmer et al., 1997). In all of the studies reported, the starting materials are titanium isopropoxide and silicon ethoxide in alkaline aqueous medium (Anthony et al., 1994; Balmer et al., 1997).

As indicated previously, CST is a multi-component material for which details of composition and structure are not well defined. The commercial material, IE-911, includes two proprietary components, a dopant and a binder (Savannah River Site, 1999; W. Wilmarth, November 21–22, 1999, personal communication), which has further limited the ability of this committee to assess its physical, chemical, and mineralogical characteristics. Anthony et al. (1994) reported that hydrothermal synthesis produced a material, TAM-5, that was fine grained and consisted of >99 percent phase-pure aggregates of small crystalline particles with an average size <100 nm. The variants of TAM-5 and IE-910 have been reported as the same materials and described as an "... inorganic ion exchanger with a well-defined crystal structure ..." (Zheng, Gu, and Anthony, 1995). Commercial manufacture produces CST in the form of extremely fine particles (IE-910), which must be agglomerated into beads (IE-911) for the intended use.

Su, Balmer, and Bunker (1997) conducted studies of the thermal stability of cesium loaded TAM-5 to 1000 °C, and they concluded that the material exists, as binary mixtures of phases, not phase-pure materials. This observation is consistent with information presented to the committee (W. Wilmarth, November 21–22, 1999, personal communication) suggesting that there are two crystalline phases with known structures, at least one crystalline phase with unknown structure, and a non-crystalline fourth phase in commercial CST (IE-911). Su, Balmer, and Bunker (1997) reported that the two crystalline materials are "known sodium- and sodium-titanium-containing phases that also contain a proprietary component." Six hydrous sodium titanosilicate compounds are reported in the open literature for this chemical system, including three mineral phases (Sokolova et al., 1985; Khalilov, 1965; Dadchov and Harrison, 1997) and three synthetic phases. Two of the reported synthetic materials are hydrous sodium titanosilicate phases (Poojary, Cahill, and Clearfield, 1994; Clearfield, Poojary, and Bortun, 1996) described as having ion exchange properties.

ISSUES TO BE ADDRESSED FOR CST

Several key issues have been identified (Savannah River Site, 1999) by SRS that would need to be resolved if CST ion exchange were to be implemented at SRS. These include the interface with MST processing for strontium and actinide removal, effect of variation in waste-feed composition, column design parameters, influence of glass formulation with higher titanium loadings on waste form performance, stability of the loaded material as a function of temperature, reproducibility of the manufactured CST, and reaction of CST in alkaline solution to form new solid phases. The committee's interim report highlighted the last two concerns, specifically noting that they would need to be resolved before the CST process could be deployed. In addition to these concerns, the large column sizes proposed for CST processing (discussed later) will result in high radiation fields with potential problems from radiolytic gas generation and heating. Work is ongoing in each of these areas and important findings are described below.

Potential problems with MST processing are discussed in Chapter 3. That treatment is intended to remove strontium and actinide ions, but it is not entirely clear how that step will interface with the CST process. It has been reported (National Research Council, 1996) that silicotitanates are selective for both cesium and strontium, although the committee has not received information about its selectivity for actinides. Consequently, the interplay between the sequential steps of supernate treatment with MST and CST could allow some flexibility in the overall processing.

Using the existing models developed at Texas A&M, the IE-911 variant of CST was evaluated (Savannah River Site, 1999) at SRS. The goal of this effort was to "... determine the performance of the CST in column applications using SRS simulated waste to determine agreement with computer modeling." The results of two column tests indicated agreement with the Texas A&M modeling, although the assumption of a 30 percent reduced cesium exchange capacity was required to model the higher flow rate (Wilmarth et al., 1999). Further evaluation during the "decision phase" of the study resulted in an unexpected increase in experimental breakthrough in the column experiment (Walker, 1998). Conflicting interpretations of the loss of exchange capacity have also been reported (McCabe, 1997; Walker et al., 1998).

Uncertainty in the nature of the ion exchange material is illustrated by a report (Walker et al., 1998) on the results for both batch and column testing for IE-911 from "batch 2" using SRS supernatant from Tank 22H. If CST continues to be an option for processing the SRS waste, SRS plans to engage ORNL and UOP to examine revised manufacturing processes to improve and ensure consistency of the CST product. Cross-laboratory comparisons of material performance are also currently underway. The Savannah River Technology Center (SRTC) also plans to "perform an evaluation of various tank wastes during the next several years," the purpose of which is to "catalogue the cesium removal efficiencies of the currently marketed CST versus the chemical compositions of F- and H-area wastes" (Savannah River Site, 1999).

Reasons for the concern about the stability of loaded ion exchange material are illustrated by a batch test of CST IE-911. The CST exposed "... to salt solution at elevated temperature ... for long duration resulted in a loss of cesium sorption capability. When the slurry cooled to room temperature, cesium did not adsorb to the IE-911." The reason for this irreversible behavior is not understood. Problems of stability in caustic solution will be addressed by long-term exposure testing to be performed by ORNL. SRTC believes that desorption may result from the formation of sodium aluminosilicate, and it plans to conduct experiments on the effects of silicon and aluminum. In addition, UOP has plans to modify its manufacturing procedures to avoid the presence of other materials.

In general it is expected that inorganic ion exchange materials, such as CST, would exhibit high stability to radiation. However, the current plant design places very stringent requirements on both the material and the process. The current column design is based on an assumed 2.5–5.8 MCi of cesium-137 in a 5-foot (1.5-meter) diameter by 16-foot (4.9-meter) long column. A 5.8 MCi loading corresponds to a radiation field of 0.66 Mrad/hr within the column, and a total dose of 1.4×10^9 rad, assuming that the column stays loaded for 3 months (Jones, 2000a). The column design and size were determined on the basis of the required processing rates and decontamination factor (Jones, 2000a). Radiation stability tests in "various waste simulants" have been conducted at the Sandia National Laboratory on IE-910 and IE-911 up to 1.2×10^9 rads at 1.4 Mrad/hr at room temperature with no observed decrease in performance. Although this provides information on radiation stability, no tests on actual wastes that would assess column flow and the effects of heat loading have been performed (or are planned) at levels close to plant design.

Work has been done to investigate potential gas generation due to radiolysis in the high fields expected to be present in the columns. Irradiation tests on IE-911 slurries have shown that hydrogen, oxygen, and nitrous oxide are produced. Gas generation tests were conducted to provide information on how gases are retained and released in the column bed. Though some issues have been identified with respect to hydrogen generation and hydrogen peroxide poisoning, no major poisoning problems were identified, and further tests are planned. SRS plans to conduct additional tests to examine cesium removal performance in the presence of gas generation, but the specific tests to be conducted have not yet been decided. Possible tests include use of hydrogen peroxide for a non-radioactive test and the use of ORNL's High Flux Isotope Reactor for a radiation exposure test, although the applicability of such reactor-based tests has not been established.

The non-elutable nature of CTS used in this ion-exchange process would necessitate that the CST as well as the cesium salts be incorporated into borosilicate glass at the DWPF. This introduces questions about the effect of higher titanium content on the stability of the glass. SRS performed a systematic survey of ten experimental glass compositions (Andrews and Workman, 1997). CST was varied from 5 to 22 weight percent, sludge content was varied from 25 to 35 weight percent, and glass frit made up the difference in the bulk composition. This survey established that a glass formulation with 5 percent CST and 28 percent sludge, formed a durable glass with viscosity and liquidus properties within the acceptable processing ranges. Edwards, Harbour, and Workman (1999) identified the ineffectiveness of modeling homogeneity. They concluded that the viscosity model used in previous studies appears to be acceptable and, in fact, over-predicts the measured viscosities. SRS plans to conduct studies on crystal growth kinetics, the effects of variation of chemical constituents, liquidus temperature bounds, and phase separation of the CST.

ANALYSIS

It is clear from the literature that CST contains multiple phases, including proprietary material that is introduced in the manufacturing process. The variability in composition of CST poses a major impediment to interpretation of the large body of data in the literature, yet these data form the basis for acceptance of CST in the proposed application. Additional uncertainty is introduced by variability in the manufacture and production of CST. In comparison to the samples studied and reported in the peer-reviewed literature, material produced commercially has been prepared by different set of chemical reactions. The form of CST for use with tank wastes is described in the literature as a sodium salt, but it is manufactured and distributed by UOP in a protonated form at pH 3; pretreatment is needed to convert it to the sodium form (Figure 5.1). The pretreatment may be contributing to the currently observed CST instability. A further problem exists in labeling these materials, which sometimes carry the same designations even when derived from different processes. The problem has been exacerbated by inconsistencies in protocols that have been used at SRS.

It is not uncommon to experience difficulties in scaling up chemical engineering processes from bench-scale to pilot-scale to full-scale operations, and this may explain some of the problems observed at SRS. But other difficulties may be related to the composition—or variability in composition—of the CST. It was reported at the committee meeting in November (W. Wilmarth, November 21–22, 1999, personal communication) that several batches of the CST were recalled because of quality acceptance and quality control difficulties. These difficulties could readily account for the loss of capacity. It also appears that the proprietary binder has not yet been adequately evaluated for its solubility at the high pH found in SRS wastes; the reaction of commercial CST in alkaline solution to form new solid phases may be responsible for observed decreases in column performance. Some of the observed difficulties may be the result of lot-to-lot variations or perhaps the change from titanium isopropoxide to chloride and then to nitrate precursors.

To resolve these questions, it will be essential to establish the relationship between the composition of CST and its performance in the ion exchange process. An assessment of the source of the inconsistency in the performance of CST must be a shared responsibility between the manufacturer and SRS. For example, it is possible that some of the observations of poor CST performance may not be caused by variations in the manufactured material but could instead result from failure on the part of SRS to prescribe and follow a uniform pretreatment protocol for the CST. It is appropriate that SRS seek assistance from Sandia National Laboratory to develop an understanding of the mechanism that governs the CST function. Since the modeling effort relies very heavily upon the studies from Texas A&M University, and in light of the observed discrepancies between the performance of CST and the modeling of the performance, a careful re-examination of the assumptions that went into the modeling is in order.

FINDINGS AND CONCLUSIONS

From the preceding discussion, it is clear that a number of issues must be addressed before CST ion exchange can become a viable technology for removing cesium from SRS HLW. A review of the available literature on CST suggests several categories of problems:

1) The issues identified with reproducibility of CST performance (for example, variability in exchange capacity and possible dissolution/reprecipitation of the proprietary components) are not well understood. These have been tentatively traced to manufacturing variability at production scale, but they could also be a consequence of the variations in pretreatment of the material at SRS or in the testing used to characterize the materials.

2) The column design for the CST process was based on typical parameters such as throughput and decontamination factors, but do not appear to be optimized for thermal loadings and expected radiation fields. The combination of particle size of the ion exchange material and the high thermal and radiation loading causes the committee to have concerns with the large-column concept and the use of this concept to drive future development.

3) The possible problems of radiolytic gas generation have not been resolved; formation of gas inside the columns could disrupt the flow of liquid and reduce the efficiency of the ion exchange process.

4) The potential effects of incorporating loaded CST into the borosilicate glass stream are not well understood. But the cesium loading capacity of the CST columns will affect the quantity of material and titanium content of the glass feed (see Chapter 8).

RECOMMENDATIONS

1) Efforts should be made—in conjunction with the manufacturer of CST—to ensure that a consistent and reproducible material is obtained for use in the ion exchange process. The basis for consistency and reproducibility should include clear and relevant acceptance criteria for the manufactured material.

2) Uniform CST pretreatment and testing protocols should be developed.

3) Column ion exchange design (and other possible processing streams using CST, such as a slurry) should be reevaluated and optimized, as should the elutable ion exchange resin.

4) An R&D effort should be undertaken to determine the performance characteristics of the standard material. This should include study of temperature effects, radiation effects, capacity, flow, gas generation, and other effects, including time, that are important for process and column operation and design.

6

Caustic Side Solvent-Extraction Process

The goal of the proposed solvent-extraction process is to extract cesium ions from the aqueous supernate into a second (nonaqueous) solvent, thereby reducing the radionuclide content of the supernate to a sufficiently low level that it could be directed to the saltstone facility for disposal as a low-level waste. A schematic description of the process is shown in Figure 6.1.

A typical solvent-extraction process for hydrometallurgical operations includes several steps. First, the aqueous feed stream is contacted with an organic solvent that is immiscible with the aqueous phase. During this contact—called extraction—one or more target components undergo transfer from the aqueous stream to the organic solvent, now called the *extract*, while other components remain in the aqueous phase. Subsequently, the loaded solvent is sent to a scrubbing and stripping operation in which the impurities are captured and returned to the feed stream. The target component(s) are transferred to a separate aqueous stream. Scrubbing and stripping both work through extraction in the opposite direction, transferring components from the organic solvent back into an aqueous solution that may differ from the original aqueous feed in pH or concentrations of other ionic species. The stripped organic solvent can then be recycled, and the target components can be treated by whatever process is appropriate. In this fashion a process can be designed to allow continuous extraction of the target components from an input stream and produce an output stream called the *raffinate* that is nearly free of the target components.

For industrial and other large-scale processes (i.e., larger than laboratory scale), extraction typically is conducted in a continuous process rather than the batch process depicted schematically in Figure 6.1. The continuous extraction process—a countercurrent extraction—often employs a vertical column, with one phase moving up the column and the second (more dense) phase moving down the column. In such an arrangement, multiple stages can be employed, and very high removals of target component(s) can often

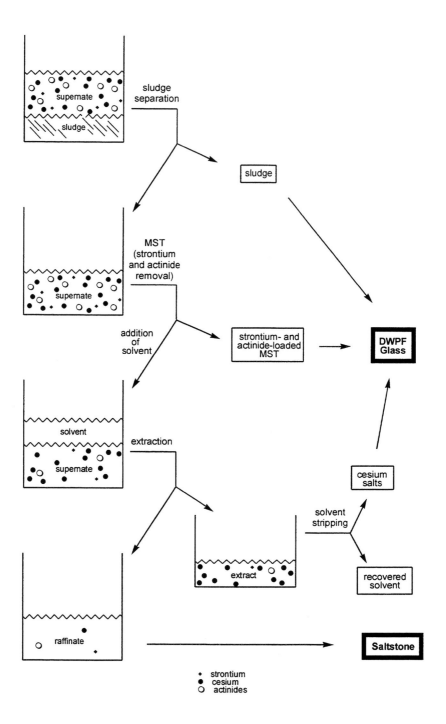

FIGURE 6.1 Schematic flow sheet for radionuclide removal from high-level tank waste at Savannah River using caustic side solvent extraction.

be attained. In the process proposed by the Savannah River Site (SRS), countercurrent extraction would be accomplished with the use of a version of the *centrifugal contactors* developed and operated at SRS since 1956, in which efficient mixing is accomplished by rotating the vessel containing the aqueous and organic phases and separating them by centrifugal force.

Solvent extraction has had a long history of successful use in the nuclear industry for operations such as spent fuel reprocessing and plutonium recovery. This experience includes exposing various organic solvents to high-radiation fields without experiencing catastrophic degradation rates. Previous experience has been mainly with aqueous feed streams that are acidic, and the key to all solvent extraction operations usually centers on the selective transfer of specific nuclides into the organic solvent.

DESCRIPTION OF THE PROCESS

According to Savannah River, the solvent-extraction process for cesium removal must remove approximately 99.998 percent of the cesium (which will require a decontamination factor, or DF, of approximately 50,000) from an aqueous, tank-waste feed stream to meet the regulatory requirements for saltstone. In the present case, the aqueous solution (produced by solvent stripping) that contains the target components—cesium ions—would be directed to the Defense Waste Processing Facility (DWPF). The cesium-free raffinate would be sent to the saltstone facility.

The use of solvent extraction for removal of cesium from aqueous tank waste at the SRS is made more difficult than from waste at other locations because of two factors. First, the SRS feed stream is highly alkaline, which renders problematic the use of the organic solvents used in many other nuclear-industry separations that employ an acidic feed. And second, the solvent must be highly selective for cesium ions as compared especially to sodium and potassium ions. All three are chemically similar, but their concentrations vary widely. The concentration of sodium ions in the supernate is typically higher than that of cesium by four orders of magnitude. The potassium ion concentration, while lower, is for many of the tanks still two orders of magnitude higher than that of cesium. Consequently, a non-selective solvent would also extract the vast fraction of the sodium and potassium ions, thereby dramatically—and unacceptably—increasing the quantity of solids to be processed in the DWPF.

Several potential challenges might be expected in developing a solvent system that is highly selective for cesium and also is thermally, chemically and radiologically stable. In addition, there is concern that insoluble material may build up at the interface between the aqueous and organic phases, a known processing issue in industrial solvent extraction that may be exacerbated by the high alkalinity of the supernate. Nevertheless, there are several potential advantages for using such a process at SRS. These include the following:

- As a general method of separation, solvent extraction is a mature technology and has been used in the nuclear industry for over 50 years (although primarily with acidic media).

- All processing occurs in liquid phases—an inherently simpler operation than performing a separation in a solids-containing stream—and no additional solids are formed. This contrasts with other alternatives for cesium removal: the small tank TPB process would generate additional solids that would need to be catalytically destroyed in subsequent processing, and the loaded inorganic ion-exchange material from the CST process would need to be sent to the DWPF.

- The process would interface well with the existing high-level waste facility.

- Solvent extraction has a wide range of operability as illustrated by its use in spent fuel reprocessing with primary feeds that have main constituents varying from aluminum to zirconium to uranium.

The proposed process uses a chemically complex, cesium-selective solvent system in which the cesium selective component is a calixarene crown ether, in which two crown ethers bridge the calix[4]arene component. The specific material proposed for use at SRS is called BoBCalixC6, for short. Other components in the solvent system include a proprietary modifier, a tertiary amine, and a hydrocarbon diluent. The calixarene crown ether appears to work through a combination of effects to generate a cavity that preferably incorporates cesium relative to other ions. The general structure shown in Figure 6.2 shows that there are two cavities, each of which is defined by a polyether macrocyclic ring containing six oxygen atoms capable of solvating the included ion. In addition, each cavity is further restricted in shape and size by two aromatic rings in the central belt of the molecule to which the macrocyclic ether rings are anchored.

Thorough demonstration of the process on an experimental scale that can confirm whether it should be seriously considered as a primary method for cesium removal from the SRS high-level tank waste has only been partially accomplished. The solvent extraction step has been demonstrated in a batch operation at Oak Ridge National Laboratory (Delmau et al., 1999)—but not continuously—using simulated feed streams. Originally, there was a problem with stripping of the cesium from the complex, but the source of the problem has since been identified. It can be solved by the addition of a tertiary amine to the solvent mix. A patent application (Moyer et al., 1999) has been submitted on this methodology.

Some continuous solvent-extraction studies have been performed at Argonne National Laboratory (ANL) on a simulated feed stream using laboratory centrifugal contactors (Leonard et al., 1999). These tests seem to have gone reasonably well, and the required separation performance was approached closely. Highly efficient, low-residence-time contactors such as those tested are absolutely essential in order to minimize the inventory of the very expensive solvent mixture. Scaled up versions of these contactors were

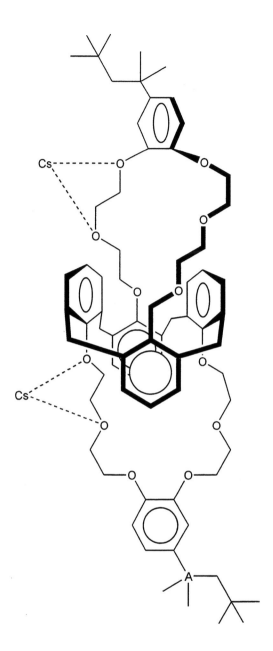

FIGURE 7.2 The calixarene crown ether (BobCalixC6) proposed for use in cesium-selective extractions. Two potential cesium-coordinating sites are indicated.

installed for other purposes in the canyons at SRS and have been operating successfully for a number of years.

The latest complexing agent/diluent/modifier blend appears to be reasonably stable in the presence of radiation and the chemical environment it would encounter in the extraction process (Delmau et al., 1999). However, this recently-developed solvent system was not used in the centrifugal contactor studies at ANL described above.

ANALYSIS, FINDINGS AND CONCLUSIONS

One chief finding has been that the technical maturity of this particular solvent-extraction process lags significantly behind the technical maturities of its two competing processes. Most of the questions to be answered, however, might be called operating concerns. They can only be answered through extensive testing of the complete process. Specific questions to be addressed include:

- Can the solvent be successfully cleaned after it becomes contaminated?
- Will "crud" build up at the interface between the two phases (a common experience in solvent-extraction processes) and thereby limit the simple separation of these phases? If encountered during processing, this could cause an unacceptable loss of solvent as well as reduce the cesium-removal efficiency.
- How will changes in the feed composition, pH, temperature, ratios of solvent constituents, and other operating variables affect the separation efficiency and capacity of the system? Will the extraction, stripping, and scrubbing operations be so delicate as to require especially sophisticated control systems to maintain steady-state, high DF operation?
- Will various ions from the tank waste or other materials in the feed stream tend to build up in the solvent phase and reduce separation efficiency or reduce the carrying capacity of the calixarene crown ether for cesium?
- Will a reliable supply (and supplier) become available for the large quantities of the calixarene crown ether that would be needed?

In spite of these concerns, the potential advantages offered by the solvent-extraction process lead the committee to conclude that it is too early to remove it from consideration and that it should remain on the list of potentially viable options.

RECOMMENDATIONS

1) "Cold" demonstration on a modest scale of the solvent extraction process should be made as soon as possible. The objectives of the demonstration should be to show that the process is adequately robust. In particular, it would need to (i) identify and solve problems such as those mentioned in the previous section and (ii) show that the required DF could be achieved for different feed compositions representative of the range expected for processing of the SRS tank waste. Other important engineering variables should also be investigated. Especially important would be the identification of "show stoppers" which could eliminate solvent extraction from consideration. During this stage, laboratory work should continue on demonstrating thermal, chemical, and radiolytic stabilities of all components of the solvent. Degradation pathways and products should be identified and these products tested for adverse effects. From this work, realistic solvent makeup volume requirements should be established.

2) Design of a hot laboratory demonstration process—using real tank waste—on a scale sufficient to define the final process should begin immediately. If solvent extraction remains in contention as a possible alternative, implementation of the hot demonstration should begin as soon as a high degree of confidence in the feasibility of the process is achieved from the cold demonstration and solvent-stability tests.

3) Work should begin immediately on defining the production capability and economics for commercial quantities of the calixarene crown ether. Purity and other specifications relevant to performance of the solvent system should be developed. Suppliers should demonstrate their ability to produce batches of solvent components that conform to specifications.

7

Direct Grout Option

The direct grout option was included by the Savannah River Site (SRS) as one of the four short-listed alternative processes primarily because of simplicity, high throughput capacity, lower cost, and extensive operational experience (see Chapter 2). This option was not selected as either the primary or backup option, however, owing to perceived regulatory, political, and public acceptance difficulties. Although this option was considered as "off the table" by SRS as the committee began its review, the committee nevertheless considered the option in its information-gathering and deliberation activities, primarily because the technical maturity of this option makes it a potential approach should all of the other cesium separation options prove technically intractable or economically undesirable. Concerning the technical maturity of the direct grout option, the committee understands that SRS has never run a large volume of grout mixture with the salt supernate. The objective of this chapter is to provide a brief review and analysis of the direct grout option and to provide recommendations on possible actions.

PROCESS DETAILS

The direct grout option differs from the other three options (crystalline silicotitanate ion exchange, caustic side solvent extraction, and small-tank tetraphenylborate precipitation) in that only the "front-end" MST treatment is applied to the high-level tank waste. Cesium is not separated from the supernate and is instead sent directly in the salt supernate to the saltstone facility.

This option is referred to as "direct grout" because the salt supernate from the strontium and actinide removal process is incorporated directly into grout with a minimal amount of pre-processing. Indeed, of the four short-listed options considered by SRS, the direct grout option involves the least amount of processing and does not produce a cesium waste stream to the Defense Waste Processing Facility (DWPF). Rather, the cesium remains in the salt supernate and is stabilized by blending it into a grout mixture that is disposed of in concrete vaults located above the current water table in an onsite shallow land-disposal facility known as the Z-Area saltstone[1] disposal facility (see Figure 1.1). After emplacement of the grout, the vaults are covered with nonradioactive grout in preparation for closure—burial under an impermeable cap.

A schematic process flowsheet for this option is shown in Figure 7.1 and comprises the following steps:

1) high-level waste in the tanks is treated with monosodium titanate (MST) to remove strontium and actinides (see Chapter 3 for details of this process);

2) slurry is then processed through a continuous filter to separate the strontium- and actinide-bearing MST solid phase (slurry) from a cesium-bearing salt supernate solution;

3) salt supernate solution is transferred to the onsite Z-Area saltstone production facility, where it is blended into a grout mixture and poured into disposal vaults located at the saltstone disposal facility; and

4) the MST solid phase (slurry) is transferred for immobilization in glass at the DWPF.

In the salt supernate solution to be processed and disposed of in the saltstone production and disposal facilities for the direct grout option, significantly higher concentrations of cesium isotopes are expected than would be handled for the in-tank precipitation (ITP) process. A comparison of concentrations of some materials in the reference salt supernate waste stream used as a basis for saltstone facility operations using the planned ITP process performance parameters (Martin Marietta Energy Systems, Inc., et al., 1992; Stevens, 1999) and anticipated concentrations in a direct grout option supernate stream (Stevens, 1999; Beck et al., 1998) are shown in Table 7.1. A higher concentration of cesium-135 exists in the direct grout waste stream than in the reference salt supernate stream. Cesium-135, with a half-life of 2.3×10^6 years, can be expected to be a contributor to radiation doses for long times following disposal. Cesium-137 also has a similarly higher concentration in the direct grout waste stream than in the reference waste stream. Although having a short half-life of 30 years, cesium-137 is present in such quantities that it will remain a major contributor to the dose for as long as 15 to 20 half-lives (450 to 600 years). Other radioisotopes of concern in long-term safety are not significantly higher in the direct grout waste stream than in the reference waste stream, although actinides in grout will present a long-term problem in meeting safety criteria. The salt supernate may also contain trace levels of hazardous metals (arsenic, barium, cadmium, chromium, lead, mercury, selenium, and silver), but Jones (2000a) notes that these metals will react with the slag components of the saltstone grout mixture to generate a reaction product that is insoluble. Other, more mobile toxic

---

[1] The term "saltstone" is applied to the grout-salt supernate mixture (a blend of fly ash, slag, and portland cement with the salt supernate) after it is poured into the concrete vaults and cured.

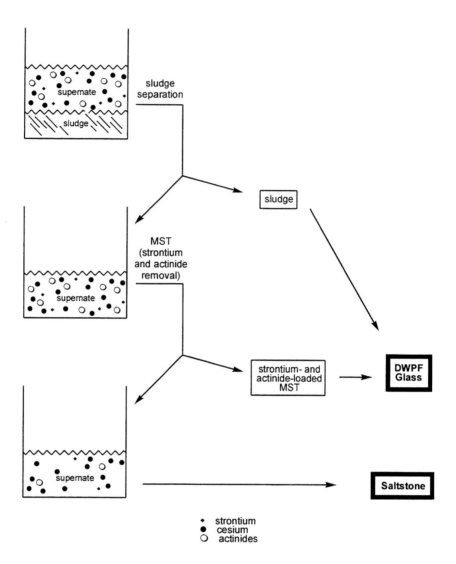

FIGURE 7.1 Schematic flow sheet for processing high-level tank waste at Savannah River using the direct grout option.

chemical constituents (primarily nitrates and nitrites) are expected to be present in similar concentrations to those evaluated and found acceptable by Martin Marietta Energy Systems, Inc., et al. (1992). Therefore, they are not expected to present a significant safety issue.

Jones (2000a) noted that the mixing of the direct grout supernate waste stream with the grout-making materials results in a dilution of the salt supernate by a factor of 1.8. Thus, the cesium-137 in the grout can be calculated to be about 2.2×10^8 nCi/L.

TABLE 7.1 Comparison of the Reference Salt Supernate Waste Stream, Direct Grout Salt Supernate Waste Stream, and Current State Permit Limits for the Saltstone Facility

Waste Component (half life, years)	Reference Salt Supernate Stream[a]	Direct Grout Supernate Stream[b]	Current State Permit Limits[c]
Sodium salts (nitrate, nitrite, sulfate, aluminate, phosphate, carbonate) and NaOH	4.5-5.0 molar	5.8 molar	no limits
Cs-135 (2.3×10^6)	5.3×10^{-2} nCi/L	8.0×10^2 nCi/L	no limits
Cs-137 (30)	2.7×10^4 nCi/L	4.0×10^8 nCi/L	1×10^6 nCi/L
Sn-126 (10^5)	1.8×10^2 nCi/L	2.2×10^2 nCi/L	no limits
Tc-99 (2.1×10^5)	8.9×10^4 nCi/L	1.1×10^5 nCi/L	3×10^5 nCi/L
Total alpha activity	0.4 nCi/g (4.8×10^2 nCi/L)[d]	limit <20 nCi/g	10 nCi/g

[a] Nominal concentrations in salt supernate stream generated from the planned (and now abandoned) in-tank precipitation process and treatment with MST (Martin Marietta Energy Systems, Inc., et al., 1992, Tables 2.6-1 and 2.6-2, nominal blend; Stevens, 1999, p. 9).

[b] Nominal concentrations in salt supernate stream that has not been treated with the in-tank precipitation process but has been treated with MST (Stevens, 1999, p. 9). Limit for total alpha activity in the salt supernate set by SRS at <20 nCi/g to meet the limit of 10 nCi/g for the saltstone disposal facility vaults (Jones, 2000b).

[c] Limits from the wastewater permit issued by the South Carolina Department of Health and Environmental Control (SCDHEC). For radionuclides, SCDHEC requires that the average saltstone concentration not exceed the limits for Class A low-level waste as defined by the U.S. Nuclear Regulatory Commission (USNRC) in Code of Federal Regulations, Title 10, Part 61.55 (10 CFR 61.55).

[d] Value in nCi/L from Martin Marietta Energy Systems, Inc., et al. (1992, Table 2.6-2); value in nCi/g = nCi/L ÷ 1230 g/L (density of reference salt solution) (J. Sessions, Westinghouse Savannah River Company, personal communication, May 23, 2000).

OBSTACLES TO SUCCESSFUL IMPLEMENTATION

The SRS views the primary obstacles to success of the direct grout option to be potentially major regulatory and political ones as opposed to technical problems (Westinghouse Savannah River Company, 1999a). Successful implementation of this option would depend on SRS's ability to surmount, at a minimum, the following two challenges:

1) the ability to obtain an incidental waste declaration for the direct grout supernate (Appendix D); and

2) the ability to obtain a permit from the State of South Carolina Department of Health and Environmental Control (SCDHEC) to allow higher concentrations of radionuclides to be disposed of in the saltstone disposal facility.

Surmounting these challenges is viewed by SRS as highly uncertain (Westinghouse Savannah River Company, 1999a).

Waste resulting from reprocessing spent nuclear fuel that is determined to be *incidental to reprocessing* (Appendix D) can be managed as transuranic (TRU) or low-level waste, depending of the radionuclide composition of the waste. If the high-level waste at SRS were to be declared as incidental and to be processed to remove strontium and actinides (for example, by using MST as shown in Figure 7.1), the supernate could then be managed as low-level waste. Incidental waste must meet the following three criteria to be classified as low-level waste (U.S. Nuclear Regulatory Commission, 1999):

Criterion 1: The waste must receive processing to remove key radionuclides to the maximum extent that is technically and economically practical.

Criterion 2: The waste must be shown to be managed to meet safety requirements comparable to the performance objectives set out in 10 CFR Part 61, Subpart C.

Criterion 3: The waste must be incorporated in a solid form at concentrations that do not exceed the concentration limits for Class C commercially generated low-level waste. However, the U.S. Department of Energy (DOE) may establish alternative requirements for waste classification and characterization on a case-by-case basis.

Radionuclide concentrations for the direct grout option calculated by SRS might in some cases exceed the current state permitted limits for the saltstone facilities (see value for cesium-137 in Table 7.1). To implement this option, a change in permit would also be necessary (Stevens, 1999). It is not clear whether regulators and the public would be willing to permit onsite disposal of the waste even if it were declared as incidental. Jones (2000a) noted that, under the permit for the saltstone disposal facility issued by SCDHEC, the average concentration for saltstone must not exceed the limits currently specified by the USNRC for Class A radioactive low-level waste. The direct grout option would be intended to produce Class C low-level waste (see Table 7.2).

Finally, there may be an internal DOE barrier to the selection of this option in any but as a last-resort case. A portion of the cesium-containing waste stream to be removed from the high-level waste supernate is planned for use in DOE's program to immobilize excess weapons-grade plutonium at SRS.[2] If the direct grout option were selected, cesium (or some other highly radioactive material) for the immobilization program might have to be obtained from another source (e.g., the cesium capsules stored at the Hanford Site in Washington).

---

[2] Under current plans, plutonium will be formed into ceramic forms (puck-shaped) and placed into steel cans. These cans will in turn be placed inside DWPF canisters, surrounded with molten high-level waste glass, and sealed. The cesium from the high-level waste will be incorporated into this glass to provide a radiological barrier against theft and clandestine use of the plutonium (see National Research Council, 1999a).

TABLE 7.2 Limits for Class C Low-Level Waste

Waste Component (half life, years)	Class C Low-Level Waste Limits[c]
Sodium salts (nitrate, nitrite, sulfate, aluminate, phosphate, carbonate) and NaOH	no limits
Cs-135 (2.3×10^6)	no limits
Cs-137 (30)	4.6×10^9 nCi/L
Sn-126 (10^5)	no limits
Tc-99 (2.1×10^5)	3×10^6 nCi/L
Total alpha activity	100 nCi/g

SOURCE: Code of Federal Regulations, Title 10, Part 61, Subpart C.

ANALYSIS

The direct grout option could be implemented with currently available technologies. The SRS has already demonstrated that it can process salt supernate from the ITP process to make grout (see Chapter 4). Because cesium is in the supernate, the saltstone production facility would need to be upgraded for possible maintenance and failures of the processing and transfer systems in a highly radioactive environment, but the committee believes that the necessary shielding and remote handling technologies are available for this application. Thus, the committee sees no significant technical hurdles to the implementation of this option unless the concentrations of radionuclides and their total amount in the grout exceed the limits for Class C low-level waste. The committee understands that there remains a fundamental issue as to whether 10 CFR Part 61 can be reasonably applied to wastes that incorporate the large inventory of cesium that would be present, about a thousand times larger than what was considered in the environmental impact statement analyses for Part 61. The total cesium-137 activity in the approximately 5.2×10^5 cubic meters of grout that would be produced would be about 210 million curies[3] (Stevens, 1999, p. 9-10).

The committee views the regulatory hurdles as probably insurmountable at this time, especially with respect to the incidental waste determination. To declare the waste as incidental (see Appendix D), DOE Order 435.1[4] requires that the "waste must receive processing to remove key radionuclides *to the maximum extent that is technically and economically practical*" (italics added for emphasis). A decision to declare the waste as incidental would

---

[3] This value is inconsistent with the total amount of Cs-137 (89×10^6 Ci) in the soluble radionuclides in the tanks as reported by Fowler (2000) and shown in Table 1.2 of this report.

[4] DOE Order 435.1 (entitled *Radioactive Waste Management*) sets out the criteria and requirements used to manage and dispose of DOE-generated or DOE-controlled waste.

represent a regulatory affirmation that processing options to remove the cesium are technically and economically impractical. This DOE order requires that the evaluation process be well documented and provide clearly traceable argument that alternative approaches eliminated are not practical.

The work to date by SRS seems, at this stage, to indicate just the opposite. That is, the documentation reviewed by the committee indicates that SRS views cesium removal as practical, using the small tank tetraphenylborate (TPB) process, but the committee notes that costs associated with cesium removal are slightly more than those associated with the direct grout option (Independent Project Evaluation Team, 1999; Stevens, 1999). Regardless of the committee's opinion of the validity of those conclusions, the fact remains that, from a regulatory perspective, current documents provide evidence that it is technically practical to treat the high-level waste at SRS to remove cesium. As such, if the cesium removal process selected can be shown to be, in the words of DOE Order 435.1, "economically practical," the waste cannot be declared as incidental waste, and therefore, the direct grout option cannot be implemented under current conditions.

If, in the future, SRS is not able to implement the current primary and backup cesium processing options for technical or economic reasons, then it might be possible to make a case that the salt supernate waste should be declared as incidental and to implement the direct grout option. At that point, however, SRS might face a number of additional and possibly insurmountable hurdles to implementation. A potentially significant hurdle is safety: SRS would have to demonstrate that the waste would meet the long-term performance objectives contained in DOE Order 435.1.

Cook (1998) has argued that it is likely that the performance objectives could be met (see Appendix E), assuming that the performance assessment[5] used for the analysis is similar to that used for the Z-area saltstone performance assessment (Martin Marietta Energy Systems, Inc., et al., 1992; Westinghouse Savannah River Company, 1998b). Barring an actual analysis, however, the ability of the site to meet these performance objectives remains uncertain. The direct grout waste stream is unusually high in long-lived radionuclides for a near-surface disposal facility. In particular, the concentration of cesium-135 in the direct grout supernate waste stream represents a long-term safety concern.

Discussants at a joint National Research Council and DOE Workshop on Barrier Technologies for Environmental Management (National Research Council, 1997) identified two key problems relevant to buried waste that could impact the direct grout option: (a) insufficient knowledge of effective lifetimes for barrier materials and systems, and (b) a dearth of barrier performance monitoring data. However, the material to be interred in the saltstone disposal facility would be a processed, solidified waste, unlike the materials typically found in radioactive waste burial sites. Other factors that must be considered in analysis of the direct grout option, especially over the long

---

[5] The performance assessment estimates long-term saltstone behavior by assessing the probability and consequence of major scenarios by which radionuclides can be released to the environment.

periods of time that wastes will exist, include stability of the thermally loaded grout and the concrete vault, possible leaching of radionuclides into the surrounding soil and groundwater, variability of the climate and the water table, natural hazards such as earthquakes and floods, and institutional stability and human intrusion.

FINDINGS AND CONCLUSIONS

Based on the foregoing discussion, the committee identified three findings and conclusions with respect to the direct grout option:

1) The committee has identified a potential major obstacle to implementation of the direct grout option: the level of cesium-137 that would be incorporated in the grout is several orders of magnitude higher than the current state permit limits (see Table 7.1), based on average tank compositions (see Chapter 8). Nevertheless, this option has the fewest technical uncertainties of the four options currently under consideration by SRS.

2) The major obstacles to implementation of this option appear to be regulatory and political. First, the committee believes that it is unlikely that the direct grout waste stream could be categorized as "incidental" given that current SRS documents provide evidence that the waste is practically and economically treatable. Second, it is not clear whether regulators and the public would be willing to permit onsite disposal of this waste even if it met incidental waste criteria.

3) The direct grout waste stream is unusually high in long-lived radionuclides for a near-surface disposal facility. The ability of the site to reliably meet long-term safety performance objectives remains uncertain.

RECOMMENDATIONS

Based on these findings and conclusions, the committee offers two recommendations, directed primarily at top-level DOE managers:

1) The direct grout option should be considered as an option of "last resort." It should be used only in case all of the other cesium processing options under consideration by SRS prove to be technically or economically impracticable and after reexamining other potentially viable options that had been previously discarded. The direct grout option should not be discarded, however, until another option has been implemented successfully.

2) DOE should hold preliminary "good faith" discussions with regulators, the SCDHEC and the U.S. Nuclear Regulatory Commission, to determine whether onsite disposal of incidental waste would be possible should the other cesium processing options be found to be impractical. Given DOE's concerns about the tight schedule for the cesium separations program in light of its tank space management problems (see Chapters 1 and 8), DOE will

need to move expeditiously to implement the direct grout option should the other options fail. Therefore, these preliminary discussions should begin immediately. Before considering implementation of the direct grout option, a thorough performance assessment would have to be conducted to determine its feasibility.

8

Barriers to Implementation of
HLW Salt Processing Options

The focus of this concluding chapter is on the third charge of the committee's statement of task (see Chapter 1): "Are there significant barriers to the implementation of any of the preferred alternatives, taking into account their state of development and their ability to be integrated into the existing Savannah River Site (SRS) high-level waste (HLW) system?" Many of the research and development (R&D) barriers to implementation of salt processing alternatives were addressed in Chapters 3 through 7 and will not be repeated here. Instead, this chapter focuses on what the committee considers to be two "global" challenges for selecting and implementing a salt processing alternative: (1) systems integration, and (2) program management.

SYSTEMS INTEGRATION

HLW salt processing is a single but key component of a much larger HLW processing system (see Figure 1.2), and the option(s) selected for processing the HLW salts must be fully compatible with the other system components. The U.S. Department of Energy (DOE) recognized the importance of *systems engineering*[1] to the success of the salt processing program when it asked the committee to comment on systems integration in this third charge. Given the abbreviated schedule for this project, the committee did not have an opportunity to perform a detailed analysis of the HLW system at SRS or the potential for integration of the candidate processing options into that system. The committee did, however, gather much information on this issue through its oral and written communications with SRS staff throughout the course of this study. Much of this information is presented in Chapters 3-7 of this report. **Based on this information, the committee concludes that**

---

[1] See *Systems Analysis and Systems Engineering in Environmental Remediation Programs at the Department of Energy Hanford Site* (National Research Council, 1998) for a good discussion of systems engineering concepts.

systems integration is not adequately implemented in the HLW salt processing options program at SRS.

The committee observed that cesium, strontium, and actinide processing are being treated as individual issues rather than as components of a fully integrated engineered system. The committee did see some evidence that systems engineering approaches are being used at SRS—for example, SRS is considering the potential impacts of waste stream feeds to the Defense Waste Processing Facility (DWPF) in its evaluation of processing options. However, other essential aspects of a systems engineering approach were lacking, especially the integrated consideration of alternative flowsheets for processing the HLW salt solutions.

The following example perhaps best illustrates the committee's conclusion on this point: As the committee became better acquainted with the HLW system at SRS, the members realized that the tank wastes were more variable in chemical and radionuclide compositions than they had been led to believe initially. The compositional differences are caused by differing inputs to the tanks from reprocessing operations (see Chapter 1) over the years and by subsequent tank transfers and processing operations. In fact, the tanks contain variable quantities of sludge, saltcake, and supernate, and radionuclide concentrations can vary from tank to tank by several orders of magnitude. This variability is illustrated for some key radionuclides in Figures 8.1 to 8.3.

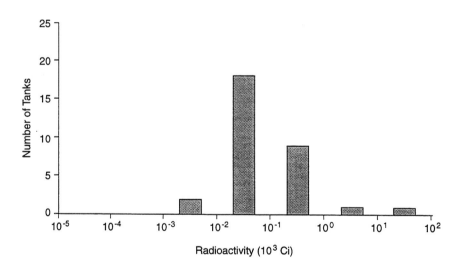

FIGURE 8.1 Histogram showing the variation of strontium-90 activity in soluble radionuclides in the high-level waste tanks at Savannah River. SOURCE: Data from Fowler (2000); see Table 1.2 of this report.

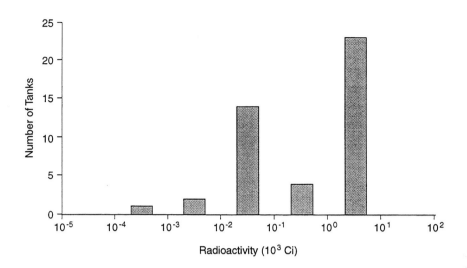

FIGURE 8.2 Histogram showing the variation of alpha activity in soluble radionuclides in the high-level waste tanks at Savannah River. SOURCE: Data from Fowler (2000); see Table 1.2 of this report.

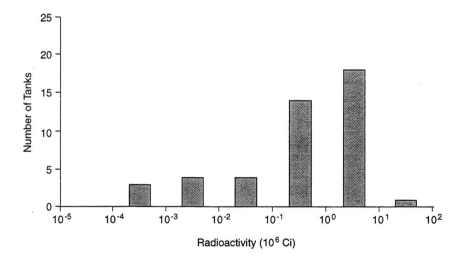

FIGURE 8.3 Histogram showing the variation of cesium-137 activity in soluble radionuclides in the high-level waste tanks at Savannah River. SOURCE: Data from Fowler (2000); see Table 1.2 of this report.

These observed intertank variabilities led the committee to pose the following questions about implementation of each salt processing option:

1) Should all tanks be subject to the same processing operations? Although the average concentrations of soluble radionuclides in the tank waste appear to be above saltstone limits (see Table 7.1), the concentrations of soluble radionuclides in some of the tanks (see Table 1.2) fall below the supernate waste stream limits for the saltstone facility—suggesting that the contents of some tanks could be sent directly to grout with little or no radionuclide removal. Thus, instead of blending tank wastes to produce a feed that might allow all tank contents to be treated by a single process, as is now planned, would it be advantageous to tailor processing based on chemical and radionuclide contents of individual tanks? For example, could tank wastes with little or no cesium be processed only to remove strontium and actinides—essentially, the direct grout option discussed in Chapter 7? Alternatively, could tank wastes with low strontium and actinide concentrations be processed only to remove cesium? Indeed, could tank wastes with low actinide, strontium, and cesium concentrations be sent directly to the saltstone facility after minor waste conditioning (e.g., filtration)?[2] Although this tailored approach might require additional regulatory approval and perhaps some facility modifications (e.g., the construction of additional waste transfer lines), it might allow the tank wastes to be processed on a faster schedule, thereby reducing costs and freeing up tank storage space for ongoing HLW processing operations. Indeed, using this approach, SRS might be able to process some tank wastes before a cesium processing option is selected and implemented, because the saltstone facility already exists. Once a cesium processing option is implemented, SRS could focus its processing efforts on the high-cesium tanks, which would produce the cesium feed stream that may be used later this decade to immobilize excess weapons plutonium.

2) Should the actinide and strontium processing step be performed prior to cesium removal? Only for the tetraphenylborate (TPB) process have advantages been presented by SRS for the MST operation as a front-end step (see Chapter 3). For other process options, the committee sees no advantages in removing these radionuclides in a front-end operation and believes that there may, in fact, be a significant disadvantage: shielding requirements are higher, thereby increasing the hazard, cost, and time of processing. The removal of cesium in a front-end step could result in much reduced radiation fields, allowing strontium and actinides to be removed in a smaller, less expensive facility. As noted previously, the MST processing step might be skipped altogether for tank wastes with low strontium and cesium concentrations.

Recommendation: SRS should implement a more fully integrated systems engineering approach for processing HLW salt solu-

---

[2] Dilution of the supernate during salt processing operations could further reduce the radionuclide concentrations shown in Table 1.2, thereby allowing more tank waste to be sent to the saltstone facility without the need for extensive radionuclide separations.

tions. **To this end, SRS should consider (1) tailoring the processing operations to tank waste contents, with the goal of reducing processing time and costs and freeing up tank space, and (2) changing the order of processing to remove radionuclides from the HLW salt solutions, with the goal of reducing processing hazards, costs, and time.**

PROGRAM MANAGEMENT

As noted in several chapters of this report, the committee has concluded that SRS generally appears to have a good understanding of the technical uncertainties that must be resolved before a HLW salt processing option can be implemented, and the committee has recommended approaches to and conduct of R&D work to resolve these uncertainties. Assuming that this R&D work is adequately funded and the appropriate people are identified to perform this work, the main barrier to the successful implementation of a salt processing option then involves two management issues: (1) ensuring that the R&D work stays properly focused on the right problems; and (2) ensuring that the salt processing program uses the information gathered by the R&D program to make appropriate selection and implementation decisions. The remainder of this chapter addresses these issues.

The experience with in-tank precipitation (see Chapter 4) illustrates how unanticipated technical "surprises" can upset even seemingly well-planned projects. Given the large volume and the chemical complexity of tank waste at SRS, such unexpected problems are possible and perhaps even likely in the future. Consequently, a primary objective of the R&D program on alternatives should be to bound performance over the range of waste and operating conditions likely to be encountered during future processing operations. This will enable engineers to design and implement a process that can accommodate such future surprises without major upsets to the high-level waste system.

The R&D program will likely be conducted at several sites across the country, and competent technical leadership will be required to ensure that this program is properly focused and coordinated. **The committee does not believe that the R&D program management should reside solely at SRS, because that site does not have the full range of technical capabilities required to direct and evaluate the required R&D work.** The committee had contact with many technical staff at SRS during the course of this review and found them, on the whole, to be a capable and dedicated group. Nevertheless, considerable experience is present at several other DOE sites (e.g., experience with cesium removal at Hanford Site and Oak Ridge National Laboratory). SRS should take greater advantage of this knowledge and experience. This work has been and will most likely continue to be carried out at a number of companies, universities, and national laboratories. The committee believes that the personnel who oversee and evaluate this work should have the similar range of technical expertise as the experts who actually perform the R&D. DOE Order 435.1 specifies that leadership and responsibility for waste management at this site resides at SRS, but close cooperation must be in place between the site and DOE Headquarters to achieve some of these objectives.

Recommendation: The committee recommends that SRS charter an external technical oversight group to guide, evaluate, and provide direction to the R&D to be carried out under this program. The group should have the same range of technical expertise as those performing the R&D work. Group members should be selected based on technical knowledge and R&D experience and should, to the extent possible, be free from professional and organizational conflicts of interest. This group should be chartered to keep the R&D focused on resolving the technical uncertainties with the processing options so that the down-selection process can proceed in a timely manner.

Similarly, the committee believes that strong and technically informed programmatic management will be needed to select and implement a HLW salt processing option at SRS. The committee understands that the salt processing program at SRS is under pressure to implement a processing option as soon as possible. Each year of program slippage has the potential to: (1) add almost $400 million to the project cost, merely from the expense of operating and maintaining the tank farms in their current condition; (2) increase the potential hazards posed by leaks to the environment from aging waste tanks; and (3) increase the likelihood that the site will run out of tank space, requiring either a curtailment of HLW operations or initiation of an expensive program to construct additional tanks.

The committee recognizes the inevitable conflicts between the objectives of taking adequate time to conduct sufficient R&D to ensure that the processing option(s) selected will do the job well, and the competing objective of completing the project in the shortest time at the most reasonable total cost. The committee also recognizes the complications introduced into R&D planning and execution by the annual budgeting system used by the U.S. government. Neither the high-risk-of-failure option ("go quickly") nor the zero-risk-of-failure option ("wait until all needed data are obtained, even if it takes forever") can be allowed to prevail. The challenge is to manage the R&D program, hot and cold demonstration testing, pilot plant operation, and process selection and implementation in ways that minimize total costs without incurring unacceptable risks, either safety or fiscal.

The successful implementation of a processing option will require strong and technically informed management of the HLW salt processing program. The committee did not see such management in place at SRS during the course of this study. Indeed, as the study progressed, it became increasingly difficult to determine who was in charge of managing this program. In the committee's view, such management is best provided by a team of experienced technical managers.

Recommendation: DOE and SRS should appoint an appropriately expert technical team with the responsibility <u>and</u> authority with adequate funding to manage the HLW salt processing program. This team should be charged with assuring that R&D and analysis of the alternate processing options are carried out as long as needed to reduce risk to an acceptable level—but no longer. The team should decide when enough information is available to allow selection or exclusion of any process, what demonstrations—hot or cold—should be conducted, what pilot plants are needed, and if so, when. The team should also be responsible for selecting the size and throughput of the production facility, since that decision also impacts both costs and risks.

References

Abley, P., and J. Halpern. 1971. Oxidation of tetraphenylborate by hexachloroiridate(IV)|. J. Chem. Soc. D (Chem. Comm.) (20):1238-1239.

Andrews, M.K, and P.J. Workman. 1997 (September 29). Glass Formulation Development and Testing for the Vitrification of DWPF HLW Sludge Coupled with Crystalline Silicotitanate (CST). Westinghouse Savannah River Company WSRC-TR-97-00312, Aiken, SC. 8pp.

Anthony, R.G., R.G. Dosch, D. Gu, and C.V. Philips. 1994. Use of silicotitanates for removing cesium and strontium from defense waste. Industrial & Engineering Chemistry Research 33(11):2702-2705.

Balmer, M.L., Y. Su, I.E. Grey, A. Santoro, R.S. Roth, Q. Huang, N. Hess, and B.C. Bunker. 1997. The structure and properties of two new siliotitanate zeolites. Material Research Society Symposium, Materials Research Society 465:449-455.

Barnes, M.J. 1990 (May 10). Decomposition of Sodium Tetraphenylborate. Westinghouse Savannah River Company WSRC-RP-90-465, Aiken, SC.

Barnes, M.J. 1992 (June 11). Sodium Tetraphenylborate Solution Stability - A Long Term Study. Westinghouse Savannah River Company WSRC-RP-92-786, Aiken, SC.

Beck, S., J.T. Carter, R.A. Dimenna, O.E. Duarte, H.H. Elder, J.R. Fowler, M.V. Gregory, T. Hang, R.A. Jacobs, P.K. Paul, R.A. Peterson, K. Rueter, P.L. Rutland, D.M. Schaffer, M.A. Shadday, F.G. Smith, III, G.A. Taylor, A.L. Whittenburg, and M.H. Wilson. 1998. Bases, Assumptions, and Results of the Flowsheet Calculations for the Short List Salt Disposition Alternatives. Westinghouse Savannah River Company WSRC-RP-98-00168, Aiken, SC.

Bianchini, C., A. Meli, F. Laschi, F. Vizza, and P. Zanello. 1989. Novel reactions of the (BPh4)- salt of the bifunctional complex ((triphos)R h(S2CO))+ (triphos = MeC(CH2PPh2)3): one-electron reduction, cleavage of B -H and B-C bonds, and metathesis of CS2-like heteroallene molecules. Inorg. Chem. 28(2):227-233.

Bonnessen, P.V., C.L. Puckett, R.V. Honeychuck, and W.H. Hersh. 1989. Catalysis of Diels-Alder reactions by low oxidation state transition-metal Lewis acids: fact and fiction. J. Am. Chem. Soc. 111(16):6070-6081.

Cho, C.S. and S.J. Uemura. 1994. Palladium-catalyzed cross-coupling of aryl and alkenyl boronic acids with alkenes via oxidative addition of a carbon-boron bond to palladium(o). Journal of Organometallic Chemistry 465(1/2):85-92.

Cho, C.S.. T. Ohe, and S.J. Uemura. 1995. Palladium(O)-catalyzed carbonylation of aryl and alkenyl boronic acids with carbon monoxide leading to esters and ketones. Transformation of a C-B bond to a C-CO bond. Journal of Organometallic Chemistry 496(2):221-226.

Clark, H.C., and K.R. Dixon. 1969. Chemistry of metal hydrides. IV. Synthesis of platinum and palladium cations (MX(CO)(R3P)2)+ and (MX(R3P)3)+. J. Am. Chem. Soc. 91(3):596-599.

Clearfield, A., D.M. Poojary, and A.I. Bortun. 1996. Structural studies on the ion-exchanged phases of a porous titanosilicate, Na2Ti2O3SiO4·2H2O. Inorganic Chemistry (21):6131-6139.

Cook, J.R. 1998. Effect of "Grout-it-all" on Saltstone Performance Assessment. Westinghouse Savannah River Company SRT-WED-98-0119, Rev. 2, Aiken, SC.

Crawford, C.L., M.J. Barnes, R.A. Peterson, W.R. Wilmarth, and M.L. Hyder. 1999. Copper-catalyzed sodium tetraphenylborate, triphenylborane, diphenylborinic acid and phenylboronic acid decomposition kinetic studies in aqueous alkaline solutions. Journal of Organometallic Chemistry 581(1/2):194-206.

Crociani, B., F. Di Bianca, L. Canovese, and P. Uguagliati. 1990. Phenylation of cationic allylpalladium(II) complexes by tetraphenylborate anion. A mechanistic study. J. Organomet. Chem. 381(1):C17-C20.

Crociani, B., F. Di Bianca, P. Uguagliati, and L. Canovese. 1991. Phenylation of cationic allyl palladium(II) complexes by tetraphenylborate. Synthesis a ALPHA-diimine olefin palladium(O) complexes and mechanistic aspects. Journal of the Chemical Society, Dalton Transaction Jan1(1):71-79.

Dadchov, M.S., and W.T.A. Harrison. 1997. JSSCB 134:409-415.

Defense Nuclear Facilities Safety Board. 1996 (August 14). Recommendation 96-1 to the Secretary of Energy pursuant to 42 U.S.C. 2286(a) (5) Atomic Energy Act of 1954, as amended. Washington, D.C.

Defense Nuclear Facilities Safety Board. 1997 (June). Savannah River Site In-Tank Precipitation Facility: Safety Implications. Technical Report DNFSB/TECH 14 (Rev 2), Washington, D.C.

Delmau, L.H., G.J. Van Berkel, P.V. Bonnesen, and B.A. Moyer. 1999 (October). Improved Performance of the Alkaline-Side CSEX Process for Cesium Extraction from Alkaline High-Level Waste Obtained by

Characterization of the Effect of Surfactant Impurities. Oak Ridge National Laboratory ORNL/TM-1999/209, Oak Ridge, TN.

Dimenna, R.A., O.E. Duarte, H.H. Elder, J.R. Fowler, R.C. Fowler, M.V. Gregory, T. Hang, R.A. Jacobs, P.K. Paul, J.A. Pike, P.L. Rutland, F.G. Smith, III, S.G. Subosits, and G.A. Taylor. 1999. Bases, Assumptions, and Results of the Flowsheet Calculations for the Decision Phase Salt Disposition Alternatives. Westinghouse Savannah River Company WSRC-RP-99-00006, Rev. 0, Aiken, SC.

Edwards, T.B., J.R. Harbour, and R.J. Workman. 1999a. Composition of Property Measurements for PHA Glasses. Westinghouse Savannah River Company WSRC-TR-99-00332. Aiken, SC.

Edwards, T.B., J.R. Harbour, and R.J. Workman. 1999b. Summary of Results for CST Glass Study: Composition and Property Measurements (U).. Westinghouse Savannah River Company WSRC-TR-99-00324. Aiken, SC.

Edwards, T.B., J.R. Harbour, and R.J. Workman. 1999c (October 4). Summary of Property Measurements from CST Glass Study. Westinghouse Savannah River Company WSRC-TR-99-00384, Revision 0, Aiken, SC. 27pp.

Eisch, J.J. and R.J. Wicsek. 1974. Rearrangements of organometallic compounds: X1. Duality of mechanism for 1,2-aryl migrations in the oxidation of tetraarylborate salts. J. Organometallic Chem. 71:C21-C24.

Ferrara, D.M., N.E. Bibler, and B.C. Ha. 1992. Radioactive Demonstration of Late Washed Precipitate Hydrolysis Process. Westinghouse Savannah River Company WSRC-RP-92-869, Aiken, SC.

Flaschka, H., and A.J. Barnard, Jr. 1960. Tetraphenylboron (TPB) as an analytical reagent. In Advances in Analytic Chemistry and Instrumentation. C.N. Reilley (ed), Vol. 1:1-105.

Fowler, J.R. 2000 (May 18). Soluble High-Level Waste Inventory Tables for the National Academy of Science. Westinghouse Savannah River Company WSRC-RP-2000-00003, Aiken, SC. 9pp.

Geske, D. H. 1959. The electrooxidation of the tetraphenylborate ion: An example of a secondary chemical reaction following the primary electrode process. J. Phys. Chem. 63:1062-1070.

Geske, D.H. 1962. Evidence for the formation of biphenyl by *intra*molecular dimerization in the electroöxidation of tetraphenylborate ion. J. Phys. Chem. 66:1743-1744.

Gu, D., L. Nguyen, C.V. Philip, M.E. Huckman, R.G. Anthony, J.E. Miller, and D.E. Trudell. 1997. Cs+ ion exchange kinetics in complex electrolyte solutions using hydrous crystalline silicotitanates. Industrial & Engineering Chemistry Research 36(12):5377-5383.

Haines, R.J., and A.L. duPreez. 1971. Extraction of a phenyl group from the tetraphenylboron anion by some .pi.-cyclopentadienyl derivatives of ruthenium. Ruthenium complex containing the tetraphenylboron anion directly bonded to the metal. J. Am. Chem. Soc. 93(11):2820-2821.

Hobbs, D.T. 2000. Response to NRC Questions, 1/11/2000. HLW-SDT-2000-00024. January 27.

Independent Project Evaluation Team. 1999 (December). Independent Assessment of the Savannah River Site High-Level Waste Salt Disposition Alternatives Evaluation — Phase IV. U.S. Department of Energy DOE/ID-10716.

Jones, R.T. 1999a (September 24). Interoffice Memorandum, R&D Expenditures for Salt Processing Alternatives. Westinghouse Savannah River Company HLW-SDYT-99-0317, Aiken, SC.

Jones, R.T. 1999b (December 14). Memorandum to File: Response to National Research Council Committee Request #1. Aiken, SC.

Jones, R.T. 2000a (January 27). Letter to Matthew Baxter-Parrott entitled "Savannah River Site High Level Waste Salt Disposition Responses to NRC Questions of 1-11-00". Westinghouse Savannah River Company HLW-SDT-2000-00024, R1, Aiken, SC.

Jones, R.T. 2000b (March 21). SRS High Level Waste Salt Disposition Document – Email Request. Letter to M. Baxter-Parrott, National Research Council, from Westinghouse Savannah River Company HLW-SDT-2000-00094, Aiken, SC.

Khalilov, A.D. 1965. DANKA 161:1409-1411.

Lee, D.D., J. F. Walker, P.A. Taylor, and D.W. Hendrickson. 1997. Cesium-removal flow studies using ion exchange. Environmental Progress 16(4):251-262.

Legzdins, P., and D.T. Martin. 1983. Organometallic nitrosyl chemistry. 20. (.eta.5-C5H5)W(NO)2BF4, a versatile organometallic electrophile. Organometallics 2(12):1785-1791.

Leonard, R.A., C. Conner, M.W. Liberatore, J. Sedlet, S.B. Aase, and G.F. Vandergrift. 1999 (August). Evaluation of an Alkaline-Side Solvent Extraction Process for Cesium Removal from SRS Tank Waste using Laboratory-Scale Centrifugal Contactors. Argonne National Laboratory ANL-99/14, Argonne, IL.

Levin, E.M., C.R. Robbins, and H.F. McMurdie. 1979. Phase Diagrams for Ceramists: Fourth Printing. American Ceramic Society, Columbus, OH.

Martin Marietta Energy Systems, Inc., EG&G Idaho, Inc., Westinghouse Company, and Westinghouse Savannah River Company. 1992 (December 18). Radiological Performance Asessment for the Z-Area Saltstone Disposal Facility. Westinghouse Savannah River Company WSRC-RP-92-1360, Aiken, SC.

McCabe, D.J. 1995 (May 18). Crystalline Silicotitanate Examination Results: Summary. Westinghouse Savannah River Company, Savannah River Technology Center WSRC-RP-94-1123, Revision 0, Aiken, SC. 15pp.

McCabe, D.J. 1997 (April 25). Examination of Crystalline Silicotitanate Applicability in Removal of Cesium from SRS High Level Waste. Westinghouse Savannah River Company WSRC-TR-97-0016, Revision 0, Aiken, SC. 15pp.

Merrell, G.B., V.C. Rogers, and M.K. Bollenbacher. 1986. The PATHRAE-RAD Performance Assessment Code for the Land Disposal of Radioactive Wastes. Rogers and Associates Engineering Corporation RAE-8511-28, Salt Lake City, UT.

Moreno-Mañas, M., M. Pérez, and R. Pleixats. 1996. Palladium-catalyzed Suzuki-type self-coupling of arylboronic acids: A mechanistic study. Journal of Organic Chemistry 61(7):2346-2351.

Moyer, B.A., P.V. Bonnesen, R.A. Sachleban, and D.J. Presley. 1999 (September 3). Solvent and Process for Extracting Cesium from Alkaline Waste Solutions, U. S. Patent Application, Serial Number 09/146,800.

National Research Council. 1996. Nuclear Wastes: Technologies for Separations and Transmutation. Committee on Separations Technology and Transmutation.Systems (STATS), National Academy Press, Washington, D.C. 571pp.

National Research Council. 1997. Barrier Technologies for Environmental Management: Summary of a Workshop. Committee on Remediation of Buried and Tank Wastes, National Academy Press, Washington, D.C.

National Research Council. 1998. Systems Analysis and Systems Engineering in Environmental Remediation Programs at the Department of Energy Hanford Site. Committee on Remediation of Buried and Tank Wastes, National Academy Press, Washington, D.C. 51pp.

National Research Council. 1999a. Interim Letter Report. Panel to Review the Spent-Fuel Standard for Disposition of Excess Weapons Plutonium, Committee on International Security and Arms Control, National Academy Press, Washington, D.C. 20pp.

National Research Council. 1999b. Interim Letter Report from the Committee on Cesium Processing Alternatives for High-Level Waste at the Savannah River Site, to E.J. Moniz, Under Secretary, U.S. Department of Energy. Washington, D.C. October 14. 27pp.

Piccolo, S.F. 1999 (November 21-22). Screening Process: Viewgraphs of presentation to National Research Council Committee on Cesium Processing Alternatives for High-Level Waste at the Savannah River Site, Augusta, GA. Westinghouse Savannah River Company HLW-SDT-99-0355, Aiken, SC. 22pp.

Poirier, M.R. 1998 (June 5). Memorandum to Steve Piccolo regarding the Evaluation of Potential Cesium Removal Technologies. SRT-WPT-98-008, Aiken, SC.

Poirier, M.R., R.D. Hunt, and C. Carlson. 1998 (May 29). Identification of Cesium Removal Technologies. WSRC-TR-98-00181. Aiken, SC.

Poojary, D.M., R.A. Cahill, and A. Clearfield. 1994. Synthesis, crystal structures, and ion-exchange properties of a novel porous titanosilicate. Chemistry of Materials 6(12):2364-2368.

Process Chemistry and Mechanisms Panel. 1996a. Summary and Results of February 8, 1996 Meeting. Aiken, SC: Westinghouse Savannah River Company.

Process Chemistry and Mechanisms Panel. 1996b. Summary and Results of February 28, 1996 Meeting. Aiken, SC: Westinghouse Savannah River Company.

Process Chemistry and Mechanisms Panel. 1996c. Summary and Results of March 6, 1996 Meeting. Aiken, SC: Westinghouse Savannah River Company.

Process Chemistry and Mechanisms Panel. 1996d. Summary and Results of March 20, 1996 Meeting. Aiken, SC: Westinghouse Savannah River Company.

Process Chemistry and Mechanisms Panel. 1996e. Summary and Results of April 27, 1996 Meeting. Aiken, SC: Westinghouse Savannah River Company.

Process Chemistry and Mechanisms Panel. 1996f. Summary and Results of May 29, 1996 Meeting. Aiken, SC: Westinghouse Savannah River Company.

Process Chemistry and Mechanisms Panel. 1996g. Summary and Results of June 27, 1996 Meeting. Aiken, SC: Westinghouse Savannah River Company.

Process Chemistry and Mechanisms Panel. 1996h. Summary and Results of July 18, 1996 Meeting. Aiken, SC: Westinghouse Savannah River Company.

Process Chemistry and Mechanisms Panel. 1996i. Summary and Results of August 21-22, 1996 Meeting. Aiken, SC: Westinghouse Savannah River Company.

Process Chemistry and Mechanisms Panel. 1996j. Summary and Results of September 12, 1996 Meeting. Aiken, SC: Westinghouse Savannah River Company.

Process Chemistry and Mechanisms Panel. 1996k. Summary and Results of October 16-17, 1996 Meeting. Aiken, SC: Westinghouse Savannah River Company.

Process Chemistry and Mechanisms Panel. 1997a. Summary and Results of January 22-23, 1997 Meeting. Aiken, SC: Westinghouse Savannah River Company.

Process Chemistry and Mechanisms Panel. 1997b. Summary and Results of April 2-3, 1997 Meeting. Aiken, SC: Westinghouse Savannah River Company.

Process Chemistry and Mechanisms Panel. 1997c. Summary and Results of June 4-5, 1997 Meeting. Aiken, SC: Westinghouse Savannah River Company.

Process Chemistry and Mechanisms Panel. 1997d. Summary and Results of September 3-4, 1997 Meeting. Aiken, SC: Westinghouse Savannah River Company.

Process Chemistry and Mechanisms Panel. 1998. Summary and Results of March 12-19, 1998 Meeting. Aiken, SC: Westinghouse Savannah River Company.

Reed, C.A., T. Mashiko, S.P. Bentley, M.E. Kastner, W.R. Scheidt, K. Spartalian, and G. Lang. 1979. The missing heme spin state and a model for cytochrome c'. The mixed S = 3/2, 5/2 intermediate spin ferric porphyrin: perchlorato(meso-tetraphenylporphinato)iron(III). Journal of the American Chemical Society 101:2948-2958.

Rogers, V.C., G.M. Sandquist, G.M. Merrell, and A. Sutherland. 1985. The PATHRAE-T Code for Analyzing Risks from radioactive Waste. Rogers and Associates Engineering Corporation RAE-8339/12-2, Salt Lake City, UT.

Rogers, V.C., and C. Hung. 1987. PATHRAE-EPA: A Low-Level Radioactive Waste Environmental Transport and Risk Assessment Code. U.S. Environmental Protection Agency EPA 520/1-87-028, Washington, D.C.

Sacconi, L., P. Dapporto, and P. Stoppioni. 1976. Synthesis and properties of cobalt and nickel complexes with the tripod ligand tris(2-diphenylarsinoethyl)amine. Structural characterization of a .sigma.-phenyl complex of nickel(II) with the same ligand. Inorg. Chem. 15(2):325-329.

Savannah River Site. 1999 (November 22). Alternative Salt Processing R&D Program Overview for "CST Non-Elutable Ion Exchange". Westinghouse Savannah River Company WSRC-RP-99-00009, Revision 1, Aiken SC.

Savannah River Site High Level Waste Salt Disposition Systems Engineering Team. 1998a (April 17). Position Paper on the Evaluation Leading to the "Initial List" of Alternatives. Aiken, SC.

Savannah River Site High Level Waste Salt Disposition Systems Engineering Team. 1998b (April 2). Pre-conceptual, Phase I Initial Design Input. Aiken, SC.

Savannah River Site High Level Waste Salt Disposition Systems Engineering Team. 1998c (April 17). HLW Salt Disposition Alternatives Identification Preconceptual Phase I, Summary Report. WSRC-RP-98-00162, Aiken, SC.

Savannah River Site High Level Waste Salt Disposition Systems Engineering Team. 1998d (June 4). Position Paper on Determination of Risk and Risk Handling Strategies for the Initial List Alternatives: Revision 1. HLW-SDT-980004. Aiken, SC.

Savannah River Site High Level Waste Salt Disposition Systems Engineering Team. 1998e (June 8). Position Paper on Dispositioning of Pro-Formas Received During Phase II: Revision 0. HLW-SDT-980014. Aiken, SC.

Savannah River Site High Level Waste Salt Disposition Systems Engineering Team. 1998f (September 17). Position Paper on the Use of Weighted Evaluation Criteria to Select the Short List of Alternatives; Revision 2. HLW-SDT-980006, Aiken, SC.

Savannah River Site High Level Waste Salt Disposition Systems Engineering Team. 1998g (October 14). Results Report on Preliminary Risk Assessment with Adjusted Risk Values: Revision 1. HLW-SDT-980015. Aiken, SC.

Savannah River Site High Level Waste Salt Disposition Systems Engineering Team. 1998h (October 29). Final Report: Recommendation Preconceptual Design (3 volumes). WSRC-RP-98-00170, Aiken, SC.

Savannah River Site High Level Waste Salt Disposition Systems Engineering Team. 1999a (June 24). HLW Salt Disposition Alternatives Identification Preconceptual Phase II Summary Report: Revision 2. WSRC-RP-98-00165. Aiken, SC.

Savannah River Site High Level Waste Salt Disposition Systems Engineering Team. 1999b (November 1). Decision Phase Final Report: Revision 0. WSRC-RP-99-00007, Aiken, SC.

Sherman, J.D. 1999. Synthetic zeolites and other microporous oxide molecular sieves. Proceedings of the National Academy of Sciences 96(7):3471-3478.

Sokolova, E.V., N.A. Yamnova, Y.K. Egorov-Tismenko, and A.P. Khomyakov. 1985. Crystal structure of a new mineral $Na_8Ti_{3.5}O_2(OH)_2(SiO_4)_4$-a polymorphous modification of natisite. Soviet Physics - Doklady 30(10):822-825.

Stevens, E. 1999 (September 13). Direct Grout Overview. Savannah River Technology Center Briefing Viewgraphs for meeting of the National Research Council Committee on Cesium Processing Alternatives, September 13-14, 1999, Augusta, GA. Savannah River Technology Center SRT-WPT-99-0011, Aiken, SC. 27pp.

Strauss, S. H. 1993. The search for larger and more weakly coordinating anions. Chem. Rev. 93(3):927-942.

Su, Y., M.L. Balmer, and B.C. Bunker. 1997. Evaluation of cesium siliotitanates as an alternative waste form. Material Research Society Symposium, Materials Research Society 465:457-464.

Turner, W.R., and P. Elving. 1965. Electrooxidation of tetraphenylborate ion at the pyrolytic graphite electrode. Anal. Chem. 37(2):207-211.

U.S. Department of Energy. 1999. Tanks Focus Area FY 1999 Annual Report. Washington, DC: Department of Energy. DOE/EM-0512.

U.S. General Accounting Office. 1999 (April). Nuclear Waste: Process to Remove Radioactive Waste from Savannah River Tanks Fails to Work. GAO/RCED-99-69. Washington, D.C.

U.S. Nuclear Regulatory Commission. 1982. 10 CFR Part 61 – Licensing Requirements for Land Disposal of Radioactive Waste," Final Rule. Federal Register 50, 38066, 1982.

U.S. Nuclear Regulatory Commission. 1999 (December 15). Classification of Savannah River Residual Tank Waste as Incidental. SECY-99-284, Washington, D.C.

U.S. Nuclear Regulatory Commission. 2000. Staff Requirements—SECY-99-0284. Classification of Savannah River Residual Tank Wastes as

Incidental. Memorandum from A.L.V. Cook to W.D. Travers, Washigington, D.C.

Walker, D., M.J. Barnes, C.L. Crawford, R.S. Swingle, R.A. Peterson, M.S. Hay, and S.D. Fink. 1996 (May 10). Decomposition of Tetraphenlyborate in Tank 48 H. Westinghouse Savannah River Company WSRC-TR-96-0113, Rev. 0, Aiken, SC.

Walker, D.D. 1998 (October 2). Modeling of crystalline siliotitanate ion exchange columns. Westinghouse Savannah River Company WSRC-TR-98-00343, Rev. 0, Aiken, SC.

Walker, D.D., W.D. King, D.P. Diprete, L.L. Tovo, D.T. Hobbs, and W.L. Wilmarth. 1998 (October 16). Cesium Removal from Simulated SRS High-Level Waste Using Crystalline Silicotitanate. Westinghouse Savannah River Company WSRC-TR-98-00344, Rev. 1, Aiken, SC. 27pp.

Walker, J.F., Jr., P.A. Taylor, and D.D. Lee. 1999. Cesium removal from high-pH, high-salt wastewater using crystalline silicotitanate sorbent. Separation Science and Technology 34(6&7):1167-1181.

Westinghouse Savannah River Company. 1998a (September 15). Bases, Assumptions, and Results of the Flowsheet Calculations for the Initial Eighteen Salt Disposition Alternatives: Revision 1. WSRC-RP-98-00166. Aiken, SC.

Westinghouse Savannah River Company. 1998b. Addendum to the Radiological Performance Assessment for the Z-Area Saltstone Disposal Facility. WSRC-RP-98-00156, Rev. 0, Aiken, SC.

Westinghouse Savannah River Company. 1999a. WSRC Briefing Packages for the NRC Committee on Cesium Processing Alternatives, September 13-14, 1999, Augusta, GA.

Westinghouse Savannah River Company. 1999b. CST-Non-Elutable Ion Exchange, Presentation to NRC Committee on September 13, 1999. SRT-LWP-99-0104, Augusta, GA.

Westinghouse Savannah River Company. 1999c. WSRC Briefing Packages for the NRC Committee on Cesium Processing Alternatives, November 21–22, 1999. HLW-SDT-99-0350 (11/23 Revision), Augusta, GA.

Wilmarth, W.R., T. Hang, J.T. Mills, V.H. Dukes, and S.D. Fink. 1999 (August 31). The Effect of Pretreatment, Superficial Velocity, and Presence of Organic Constituents on IONSIV IE-911® Column Performance. Westinghouse Savannah River Company WSRC-TR-99-00313, Aiken, SC. 25pp.

Zheng, Z., D. Gu, and R.G. Anthony. 1995. Estimation of cesium ion exchange distribution coefficients for concentrated electrolytic solutions when using crystalline silicotitanate, Industrial & Engineering Chemistry Research 34:2142-2147.

Zheng, Z., C.V. Philip, G.G. Anthony, J.L. Krumhansl, D.E. Trudell, and J.E. Miller. 1996. Ion exchange of group I metals by hydrous crystalline silicotitanates. Industrial & Engineering Chemistry Research 35(11):4246-4256.

Appendix A

Biographical Sketches of Committee Members

MILTON LEVENSON (*Chair*) is a chemical engineer with over 55 years of experience in nuclear energy and related fields. His technical experience includes work in nuclear safety, fuel cycle, water reactor technology, advanced reactor technology, remote control technology, and sodium reactor technology. His professional experience includes positions at Oak Ridge National Laboratory in research and operations, Associate Director for Energy and Environment at Argonne National Laboratory, first Director of Nuclear Power at the Electric Power Research Institute, and Vice President of Bechtel International. Mr. Levenson is the past president of the American Nuclear Society; a fellow of the American Nuclear Society and the American Institute of Chemical Engineers; and the recipient of the American Institute of Chemical Engineers' Robert E. Wilson Award. He is the author of over 150 publications and presentations and holds three U.S. patents. He received his B.Ch.E. from the University of Minnesota and was elected to the National Academy of Engineering in 1976.

GREGORY R. CHOPPIN (*Vice-Chair*) is the R.O. Lawton Distinguished Professor of Chemistry at Florida State University. Dr. Choppin conducts research in nuclear chemistry, physical chemistry of the actinides and lanthanides, environmental behavior of actinides, chemistry of the f-Elements, separation science of the f-Elements, and concentrated electrolyte solutions. During a postdoctoral period at the Lawrence Radiation Laboratory, University of California, Berkeley, he participated in the discovery of mendelevium, element 101. His research activities have been recognized by the American Chemical Society's Award in Nuclear Chemistry and Southern Chemist Award, The Manufacturing Chemists award in Chemical Education, and a Presidential Citation Award of the American Nuclear Society. He has served on numerous National Research Council committees and recently completed a six-year term as a member of the Board on Chemical Sciences and Technology. He is a member of the committee on the Electrometallurgical Treatment of EBRII Spent Fuel, the Committee on Remediation of Buried and Tank Wastes, and the Board on Radioactive

Waste Management. He received his B.S. in chemistry from Loyola University, New Orleans, his Ph.D. in chemistry from the University of Texas, Austin, and has honorary degrees from Chalmers University, Goteborg, Sweden, and from Loyola University, New Orleans.

JOHN BERCAW is the Centennial Professor of Chemistry at the California Institute of Technology. Dr. Bercaw is an expert in organometallic chemistry. His research interests include synthetic, structural, and mechanistic organotransition metal chemistry, compounds of early transition metals, and hydroxylation of alkanes by simple platinum halides in aqueous solutions. Dr. Bercaw is a former chair and Executive Committee member of the American Chemical Society's Inorganic Chemistry division. He is a fellow of the American Association for the Advancement of Science and a fellow of the American Academy of Arts and Sciences. His work has been recognized with the American Chemical Society's Award in Pure Chemistry, the Award in Organometallic Chemistry, the Award for Distinguished Service in the Advancement of Inorganic Chemistry, and the George A. Olah Award in Hydrocarbon or Petroleum Chemistry. Dr. Bercaw earned his B.S. in Chemistry from North Carolina State University and his Ph.D. in Chemistry from the University of Michigan. Dr. Bercaw was elected to the National Academy of Sciences in 1990.

DARYLE H. BUSCH is the Roy A. Roberts Distinguished Professor of Chemistry at the University of Kansas. Previously, he was a faculty member at The Ohio State University where he rose through the ranks from assistant professor (1954) to Presidential Professor (1987). His research in basic transition metal coordination chemistry fathered modern macrocyclic ligand chemistry and created the molecular template effect. He was among the founders of the subject of ligand reactions and an early researcher and proponent of bioinorganic chemistry. He first described the phenomenon called *preorganization* in 1970. His research is presently focused on homogeneous catalysis, bioinorganic chemistry, and orderly molecular entanglements. Throughout his research career, Dr. Busch has worked closely with industry and holds patents with five major industrial companies. Some of the recognitions of Dr. Busch's work include the American Chemical Society Award for Distinguished Service in Inorganic Chemistry and for Research in Inorganic Chemistry, the John C. Bailar Medal from the University of Illinois, the Dwyer Medal of The Royal Society of New South Wales, Australia, and the Izatt-Christenson International Award for Macrocyclic Chemistry. Dr. Busch has written three textbooks, and numerous book chapters, articles, and reviews. His teaching has been recognized by the University of Kansas' Louis Byrd Graduate Educator Award. He is the President of the American Chemical Society. Dr. Busch received his B.A. from Southern Illinois University and his M.S. and Ph.D. from the University of Illinois.

TERESA FRYBERGER is the Associate Laboratory Director for Applied Science and Technology at Brookhaven National Laboratory (BNL). As head of one of four science directorates at BNL, she manages and develops a diverse program in environmental sciences, energy sciences and national security, as well as applied chemistry and materials science. Prior to taking the Brookhaven position, Dr. Fryberger was a senior manager at Pacific Northwest National Laboratory (PNNL). As Senior Deputy Director of PNNL's William R. Wiley Environmental Molecular Sciences Laboratory, she was responsible for managing environmental science programs and providing strategic direction for the overall management of this National Scientific User Facility. Dr. Fryberger has managed national scientific programs at the Department of Energy (DOE), has been an associate editor at *Science*, and was a research chemist and National Research Council Postdoctoral Fellow at the National Institute for Science and Technology. She is a member of the American Chemical Society, the American Physical Society, and the American Association for the Advancement of Science. Dr. Fryberger has organized and chaired over fifty national and international meetings of professional societies, DOE, and technical organizations, and has served on numerous advisory and review committees for national laboratories, DOE, and various universities. She earned her Ph.D. in Physical Chemistry from Northwestern University and her B.S. in Chemistry from the University of Oklahoma.

GEORGE KELLER retired in 1997 from Union Carbide Corporation as a Senior Corporate Research Fellow at the Technical Center in South Charleston, West Virginia. He is presently an adjunct professor of chemical engineering at both West Virginia and Marshall Universities and is also involved with various economic-development activities involving new technologies. His prime area of expertise is separation science and technology, with a focus on adsorption, chemical complexation, absorption, distillation, and membranes. He has also been heavily involved with oxidation catalysis, thermal cracking, and energy-use minimization. He earned his B.S. in chemical engineering from Virginia Polytechnic Institute and his M.S. and Ph.D. from Pennsylvania State University. Dr. Keller was elected to the National Academy of Engineering in 1988.

MATTHEW KOZAK is a Staff Consultant at Monitor Scientific, LLC. Dr. Kozak is an expert in safety assessment and regulatory assessment for waste disposal. He is the U.S. delegate to the International Atomic Energy Agency's Coordinated Research Program on Improvement of Safety Assessment Methodologies and the Chair of the National Council on Radiation Protection Measurements Scientific Committee 87-3 on Safety Assessment of Near-Surface Radioactive Waste Disposal Facilities. Dr. Kozak regularly advises the U.S. Department of Energy and the U.S. Nuclear Regulatory Commission on safety issues. He is a member of the Health Physics Society, the American Nuclear Society, and the American Institute of Chemical Engineering. His professional honors include the Washington State

Mining and Minerals Institute fellowship and the National Academy of Sciences' Radioactive Waste Management grant in 1997. Dr. Kozak earned a B.Ch.E. from Cleveland State University and his Ph.D. in Chemical Engineering from the University of Washington, Seattle.

ALFRED P. SATTELBERGER has been a member of the technical staff at Los Alamos National Laboratory since 1984. He is currently Director of the Chemistry Division, which employs approximately 375 people and supports a variety of programs in analytical, inorganic, physical, nuclear and radiochemistry, medical radioisotopes, and nuclear physics. Dr. Sattelberger is an inorganic/organometallic chemist with research interests in actinide science, technetium chemistry, homogeneous and heterogeneous catalysis, and metal-metal multiple bonding. Prior to moving to Los Alamos, Dr. Sattelberger held a faculty position in the chemistry department at the University of Michigan. He is a past member of the Executive Committee of the Inorganic Chemistry Division of the American Chemical Society and currently serves on the Board of Directors for both the Inorganic Synthesis Corporation and the Los Alamos National Laboratory Foundation. He served as a reviewer on the General Inorganic Chemistry of the DOE Environmental Management Science Program (EMSP) merit review panel from 1996–1998 and on the National Research Council's Committee on Building an Effective EM Science Program. Dr. Sattelberger earned his B.A. in Chemistry from Rutgers College and his Ph.D. in Inorganic Chemistry from Indiana University.

BARRY E. SCHEETZ is Professor of Materials, Civil and Nuclear Engineering at the Pennsylvania State University. Dr. Scheetz is active in research areas dealing with the chemistry of cementitious systems. His activities include environmental waste management programs such as remediation of minelands by the use of industrial by-products, focusing on large-volume usages of fly-ash-based cementitious grouts. Other programs include developments of radioactive waste forms based on vitrifiable hydroceramics and sodium zirconium phosphate (NZP) structures. Professor Scheetz has serves as a participant in the Oak Ridge National Laboratory review of the Valley of the Drums and as a technical expert for the National Research Council's review of the Idaho National Environmental and Engineering Laboratory processing alternatives for calcined high-level nuclear waste. He has also served as an expert for the State Attorney General of New Mexico. Among his many accomplishments, he received a national internship from the Argonne National Laboratory in 1972, and he was a National Academy of Sciences Visiting Scholar to China in 1989. Professor Scheetz is the author of over 160 scientific publications and holds 46 United States and foreign patents. He received a B.S. in chemical education from Bloomsburg State College, a M.S. in geochemistry, and a Ph.D. in geochemistry and mineralogy from Pennsylvania State University.

MARTIN J. STEINDLER worked at Argonne National Laboratory until his retirement in 1993. His last position at Argonne was as Director of the Chemical Technology Division. Dr. Steindler's expertise is in the fields of the nuclear fuel cycle and associated chemistry, engineering, and safety, with emphasis on fission products and actinides. In addition, he has experience in the structure and management of research and development (R&D) organizations and activities. He has published more than 130 papers, patents, and reports on topics in these areas. During his career, Dr. Steindler has been a consultant to the Atomic Energy Commission, the Energy Research and Development Agency, and various Department of Energy (DOE) laboratories. He chaired both the Materials Review Board for the DOE Office of Civilian Radioactive Waste Management and the U.S. Nuclear Regulatory Commission Advisory Committee on Nuclear Waste. Dr. Steindler has served on several National Research Council committees, and currently serves on the Board on Radioactive Waste Management, the Committee on Environmental Management Technologies' Mixed Waste Committee, and the Committee on Remediation of High-Level Waste Tanks in the DOE Weapons Complex. He received B.S., M.S., and Ph.D. degrees in chemistry from the University of Chicago and a number of awards related to his work.

Appendix B

Interim Report

THE NATIONAL ACADEMIES

Advisers to the Nation on Science, Engineering, and Medicine

National Academy of Sciences
National Academy of Engineering
Institute of Medicine
National Research Council

Board on Radioactive Waste Management
National Research Council

October 14, 1999

Ernest J. Moniz
Under Secretary
U.S. Department of Energy
Washington, D.C. 20585

Dear Dr. Moniz:

The National Research Council empaneled a committee[1] at your request[2] to provide an independent technical review of alternatives for processing the high-level radioactive waste salt solutions at the Savannah River Site[3]. You requested that the National Research Council provide you with an interim report that identifies significant issues or problems with the processing alternatives before the Department of Energy (DOE) issues a draft environmental impact statement (EIS), which is planned for release in October 1999. The committee's interim report is provided in this letter. This report has been reviewed in accordance with the procedures of the National Research Council[4] and reflects a consensus of the committee.

The information used to develop this interim report was obtained from several sources. The committee reviewed published documents that describe the salt processing program at Savannah River and the screening process used to select alternative processing options[5]. The committee also held an information-gathering meeting on September 13-14, 1999 in Augusta, Georgia to receive briefings from DOE staff, Westinghouse Savannah River Company (WSRC) staff, and National Laboratory scientists[6] on the alternative processing options and current and planned research and development (R&D) activities.

The committee does not yet have enough information to fully address its statement of task[7]. However, based on the information gathered to date, the committee has reached several conclusions that it believes will be helpful to DOE in finalizing the draft EIS. These conclusions are described in the following paragraphs and are organized around the four bullets of the

---

[1] Committee on Cesium Processing Alternatives for High-Level Waste at the Savannah River Site. The roster for this committee is given in Attachment A.
[2] A copy of your letter of request to the National Research Council is given in Attachment B.
[3] An overview of the high-level waste program and the alternative processing options is provided in Attachment C.
[4] The list of report reviewers is provided in Attachment D.
[5] A list of documents received by the committee is provided in Attachment E.
[6] See Attachment F for a list of the presentations and personnel involved in the committee's first information-gathering meeting.
[7] The committee's statement of task is given in Attachment G.

Ernest J. Moniz
October 14, 1999
Page 2

committee's statement of task. The committee also offers several recommendations in the concluding paragraphs of this report.

- <u>Task 1: Was the process used to screen the alternatives technically sound and did its application result in the selection of appropriate preferred alternatives?</u>

The screening process (see Attachment C) to identify cesium removal alternatives was undertaken by the Salt Disposition Systems Engineering Team under the sponsorship of WSRC. This team was comprised of 10 members with expertise in science and engineering, operations, waste processing, and safety and regulations. The team interacted with experts throughout the DOE complex and undertook a historical review and literature survey to identify about 140 possible processes that could potentially be used to process the high-level waste salt solutions at Savannah River. These processes were grouped into an "initial list" of 18 alternative processing options, which were subsequently screened using a multi-attribute analysis to obtain a "short list" of four alternative processing options: small tank tetraphenylborate (TPB) precipitation, caustic side solvent extraction, direct disposal in grout, and crystalline silicotitanate (CST) ion exchange. This screening process has been reviewed by numerous groups, including two expert teams assembled by DOE, and has received generally favorable marks.

The committee has not yet had an opportunity to perform a detailed examination of this screening process. Therefore, a full response to this part of the task statement must be deferred to the committee's final report. However, the committee does have one comment at this time relative to this question: Given the ambitious schedule that the Department has defined for selecting and implementing a process for treating the cesium-bearing salt solutions at Savannah River—a draft EIS is to be issued in October 1999, a Record of Decision (ROD) is to be made in spring 2000, and the selected option is planned to be implemented no later than 2008[8]—a negative answer by the committee to this statement-of-task question could delay Savannah River's plans to process this waste and could markedly increase the total cost of the processing operations[9]. This question could have been asked earlier to permit more meaningful input into the screening process.

The storage of high-level liquid wastes in underground tanks, some of which are several decades old, represents a potential hazard to workers and the environment at the site and a continuing burden on U.S. taxpayers. The committee shares the Department's (and WSRC's) sense of urgency to address this hazard by removing and treating the waste as soon as safe and practical. Consequently, in addressing this part of its statement of task, the committee will be asking the question "Did the screening process lead to the identification of technically sound options for processing the waste?" The committee's initial impression is that the screening process did result in the identification of several potentially viable alternative processing options. The committee will perform a more detailed review of the overall screening process during the remainder of this study.

---

[8] It is not clear to the committee <u>how</u> this process will be implemented. The committee learned that DOE will likely issue a request for proposals (RFP) from industry to implement one of these options. However, DOE Savannah River staff were unable to provide the committee with any details on this RFP.
[9] According to WSRC staff, the operating costs of the high-level waste system at Savannah River are about $400 million per year.

Ernest J. Moniz
October 14, 1999
Page 3

  • __Task 2: Was an appropriately comprehensive set of cesium partitioning alternatives identified, and are there other alternatives that should be explored?__

Given the compressed schedule for producing this interim report, the committee has focused most of its attention on the four options that were included in the "short list" of alternatives discussed in Attachment C. The committee has not yet had the opportunity to perform a detailed review of the full list of alternatives for processing the salt solutions that were identified by WSRC through its alternatives screening process. The committee did, however, perform a cursory examination of the list of 18 alternative processing options developed by the Salt Disposition Systems Engineering Team. These options included the approaches that were known to committee members to be useful for processing cesium-bearing alkaline salt solutions. Thus, the committee's initial impression is that no major processing options were overlooked in the screening process. However, the committee will perform a more thorough review of alternative processing options for the final report.

  • __Task 3: Are there significant barriers to the implementation of any of the preferred alternatives, taking into account their state of development and their ability to be integrated into the existing Savannah River Site HLW system?__

The committee examined the four alternatives that passed the multi-attribute screening process (i.e., small tank TPB precipitation, caustic side solvent extraction, direct disposal in grout, and CST ion exchange) to assess whether there were significant barriers to implementation. The committee concluded that any of these four alternatives could probably be made to work if enough time and funding were devoted to overcoming the remaining scientific, technical, and regulatory hurdles. However, the time, cost, and technical risk of implementation of each option could vary widely because all are at different states of development.

A preliminary summary of the scientific, technical, and regulatory hurdles for each option is summarized below. For the small tank TPB precipitation, caustic side solvent extraction, and CST ion exchange processing options, the remaining hurdles are both scientific and technical in nature and include the need for obtaining a better understanding of basic chemical processes. The direct disposal in grout option appears to be technically mature but faces significant regulatory hurdles.

Small tank TPB precipitation. The small tank TPB precipitation option was developed by WSRC staff to "engineer around" the benzene production problem discovered during large tank in-tank precipitation (ITP) operations (see Attachment C). The development of this option was based on the belief by WSRC staff that they adequately understood the basic chemistry and process phenomena that led to the earlier difficulties with the ITP process.

This new option was designed both to reduce the production of benzene during processing and storage of the waste and to reduce benzene explosion hazards. As currently designed, the process will employ 57,000 liter (15,000 gallon) stainless steel reaction vessels and short processing and storage times that, taken together, allow less TPB to be used in processing and reduces the time available for radiolytic and catalytic decomposition of the TPB to form benzene. Additionally, the reaction vessels are designed to maintain a positive pressure so that the head (open) space can be blanketed with a nitrogen atmosphere to reduce explosion hazards. This design also allows for the capture and treatment, if desired, of benzene evolved during processing operations.

Ernest J. Moniz
October 14, 1999
Page 4

Although a positive pressure would allow the reaction vessels to be blanketed with inert gas, this design could promote the release of benzene and possibly of radionuclides to the environment should a leak occur. The standard practice in biological and radiological facilities is to maintain a negative pressure relative to the atmosphere so that leaks result in inflows rather than outflows. The use of positive pressure reaction vessels for this process may require additional containment and safety procedures to minimize hazards should leaks occur. As noted in Attachment C, the current design for this process includes provisions for secondary containment.

Small tank TPB precipitation appears to be the alternative preferred by WSRC for processing the cesium-bearing salt solutions at Savannah River. WSRC has over 16 years of experience with TPB and has developed a rudimentary understanding of the cesium precipitation process through R&D work done during ITP development and operations. This option appears to have the most advanced R&D and engineering development of all of the options except direct disposal in grout.

Nevertheless, the committee believes that additional R&D is required to demonstrate that this option could be successfully implemented to treat the cesium-bearing salt solutions at Savannah River. WSRC lacks an adequate understanding of the chemistry underlying the TPB decomposition process and the catalysts and catalytic reactions responsible for benzene generation. In place of such an understanding, WSRC appears to be focusing on an engineering-design solution based on untested assumptions about maximum likely benzene production and catalytic pathways. The WSRC staff who briefed the committee indicated that WSRC has a limited understanding of the mechanisms of catalysis responsible for benzene production and also that WSRC has collected little experimental data on the sources or roles of likely catalysts such as soluble transition metal complexes and dispersed palladium metal particles.

The committee believes that WSRC must obtain a better understanding of the chemistry of the TPB decomposition process before this option could be selected and deployed to treat the cesium-bearing salt solutions at Savannah River. The extreme complexity of the chemical system in the alkaline tank waste at Savannah River—which consists of more than 35 elements in a variety of phases and chemical compounds, including solid and liquid complexes—increases the likelihood that significant and unanticipated technical problems will be encountered unless benzene generation and release processes are better understood[10]. The committee believes that it would be advantageous from both time-efficiency and cost-efficiency standpoints to undertake this R&D work before this process is selected and deployed. The alternative—namely, to proceed with deployment immediately and engineer around the gaps in chemistry knowledge—carries a high technical risk and could result in a repeat of the ITP failure.

CST ion exchange. Crystalline silicotitanate is an inorganic material that has a high selectivity for cesium over other alkali metals and is thus a potentially useful ion-exchange material for cesium removal from alkaline tank waste. Although ion exchange for cesium removal is a well known technique, CST ion exchange has never been used in a large-scale nuclear waste application, and CST has never been manufactured in commercial-scale

---

[10] This conclusion was previously reported in Recommendation 96-1 from the Defense Nuclear Facilities Safety Board.

Ernest J. Moniz
October 14, 1999
Page 5

quantities. Consequently, additional R&D will be required to demonstrate this technology for cesium removal from tank waste at Savannah River.

The committee learned during its information-gathering meeting that WSRC has discovered two significant and potentially insurmountable problems with the CST ion exchange process. First, WSRC staff discovered that cesium is desorbed (i.e., released) from CST at elevated temperatures—this phenomenon was observed to occur at temperatures of 50 °C and probably operates (albeit at reduced rates) at lower temperatures, including at the planned 25 °C to 30 °C processing temperatures. This desorption process appears to be irreversible— that is, once cesium is released, it is not reabsorbed once temperatures are lowered. Second, CST appears to react with constituents in the alkaline tank waste to produce new solid phases that may be capable of plugging the ion exchange columns. Because either of these problems could lead to extended and costly shutdowns of tank waste processing operations at the site if this processing option were to be implemented, these problems must be resolved before this process can be deployed.

WSRC staff do not yet understand the chemical processes responsible for either of these phenomena, although they speculated to the committee that the tank waste reactions with CST may be due to manufacturing impurities. Additional R&D work is required to address these problems.

Caustic side solvent extraction. Solvent extraction is a mature and widely implemented technology for separating uranium and plutonium from acidic solutions (e.g., in the PUREX process), but it has never been used to treat highly alkaline wastes like the cesium-bearing salt solutions at Savannah River. There is a great deal of process experience with solvent extraction across the DOE complex, including at Savannah River. Furthermore, this "all liquid" process is highly compatible with the existing high-level waste system at the site. The process would produce a liquid cesium-bearing solution that could be sent directly to the Defense Waste Processing Plant without additional processing steps.

The caustic side solvent extraction process has been developed by scientists at the Oak Ridge National Laboratory and appears to the committee to be a potential "break-through" technology. However, caustic side solvent extraction is at an immature stage of development relative to the other three options and, although there do not appear to be any insurmountable problems with this process, additional R&D will be required to demonstrate that it could be successfully implemented to treat the cesium-bearing salt solutions at Savannah River. In particular, additional R&D needs to be done to obtain performance data under real-life conditions—for example, R&D to determine the radiation stability of the solvent system (including the stability of the cesium chelator, diluent, and modifier; see Attachment C), the ability to scrub and recycle the expensive solvents, and the ability to mitigate contaminant formation during processing. It may be possible to answer these questions relatively quickly in a small-scale pilot study under process conditions. Additionally, the cesium chelator for this system has never been manufactured in commercial quantities, so a significant scale-up of manufacturing capabilities would have to be demonstrated to ensure that reagent-grade material could be produced in quantities required for processing the cesium-bearing salt solutions at Savannah River.

Direct disposal in grout. This process is very similar to the so-called "saltstone process" that was to have been used to dispose of the salt solutions from the ITP process. As noted previously, this is a very mature technology and has already been demonstrated at the site for

Ernest J. Moniz
October 14, 1999
Page 6

less radioactive salt solutions. Some additional R&D work may be needed to develop grout that will retain cesium to satisfy regulatory requirements. In general, cesium sorption on cementitious material like Portland cement is lower than virtually all other radionuclides. Given the large inventory of cesium that would be disposed of in grout under this option, the demonstration of compliance under applicable regulations (see the next paragraph) may require the development of grout formulations with a higher cesium sorption coefficient. Additionally, engineering design would be required to develop a facility that could be remotely operated and maintained to protect workers from high radiation fields. However, the required R&D and engineering work appear to be relatively straightforward, and the committee knows of no insurmountable problems with this option.

The major hurdles with the direct grouting option are regulatory in nature[11]. The process would be intended to produce Class C low-level waste that would be disposed of at Savannah River in a land-disposal facility (see Attachment C). To be eligible for such onsite disposal, the cesium-bearing salt solutions would have to be declared to be "incidental waste" under DOE Order 435.1[12]. To be declared as incidental waste, DOE would have to demonstrate that the wastes "Have been processed, or will be processed, to remove key radionuclides to the maximum extent that is technically and economically practical" and that the waste "Will be managed to meet the safety requirements comparable to the performance objectives set out in 10 CFR Part 61, Subpart C ...." The latter criterion would require DOE to undertake a detailed performance assessment analysis to demonstrate that disposal of the grout onsite would meet the U.S. Nuclear Regulatory Commission's (USNRC's) radionuclide release criteria for land disposal facilities as well as intruder barrier requirements for Class C waste[13]. However, DOE would not be required to seek formal USNRC approval for this assessment. Additionally, DOE would likely be required to seek permits from South Carolina and/or the U.S. Environmental Protection Agency to operate this facility because it contains other regulated wastes[14]. It is not clear to the committee whether resolution of these regulatory issues would be possible under the current schedule constraints.

"Front-end" actinide and strontium removal. As noted in Attachment C, the four cesium removal options discussed above are designed to process waste streams that have been treated to remove actinides and strontium. Savannah River plans to remove these radionuclides at the "front end" of processing operations by treating the waste with monosodium titanate (MST), which sorbs actinides and strontium. To the committee's knowledge, this process has not been used elsewhere to process tank waste and therefore would be a first-of-its-kind implementation, subject to the usual problems inherent with such applications.

In fact, there do appear to be some remaining technical questions that will need to be resolved before this process could be implemented successfully at Savannah River, and the committee learned at its information-gathering meeting that WSRC appears to be pursuing these questions vigorously. In particular, MST reaction kinetics are not well understood—work

----

[11] Public acceptance may also be a significant barrier with this option, because it will entail the shallow-land disposal of isotopes that will be highly radioactive for hundreds to thousands of years.

[12] DOE O 435.1, Radioactive Waste Management, Approved July 9, 1999.

[13] Additionally, it is unclear whether the regulations 10 CFR Part 61 can even be reasonably applied to this waste—the amount of cesium to be disposed of in the saltstone would be about a thousand times larger than what was considered in the EIS analyses for Part 61.

[14] The Saltstone Facility at Savannah River, for example, operates under an permit for a landfill disposal site, even though it contains Class A levels of radioactive waste.

Ernest J. Moniz
October 14, 1999
Page 7

by WSRC staff suggests that actinide and strontium reactions with the MST may proceed more slowly than anticipated, thus requiring higher MST concentrations to achieve the required throughputs. However, titanium (a component of the MST) is incompatible with borosilicate glass, so there are limits on the amount of MST that could be used in processing operations. There are some indications that these limits could be approached or exceeded to meet actinide removal requirements for the planned disposal of the residual salt solutions in the saltstone facility. Recent work by WSRC suggests that waste dilution and waste blending (i.e., combining waste from different tanks) may be required to meet the performance requirements for this process. This additional processing will add time, cost, and technical risk to waste processing operations.

To the committee's knowledge, WSRC is not considering any alternative processes for removal of actinides and strontium from the tank waste. Thus, this process <u>must</u> work as intended for the high-level waste processing program to succeed. Additional R&D work, especially on reaction kinetics, will be required to demonstrate that this process can meet both performance and regulatory requirements. If this work identifies any insurmountable problems, then WSRC will have to find alternative processes for removing these radionuclides. The committee's initial impression is that this process can be made to work, but WSRC must get on with the pilot-scale testing to demonstrate that this process will achieve the needed throughput.

- <u>Task 4: Are the planned R&D activities, including pilot-scale testing, adequate to support implementation of a single preferred alternative?</u>

A consideration of planned R&D activities will be a major component of the committee's future work, and at this point in the study the committee only has enough information to make two general observations about ongoing and planned R&D activities. The committee's first observation is that R&D resource allocations for the four alternative processing options have been markedly inequitable. In FY99, R&D funding for the four alternative processing options totaled about $11 million—about $4.4 million for small tank TPB, $6.0 million for CST ion exchange, $0.3 million for caustic side solvent extraction, and $0.3 million for direct disposal in grout. Funding for the R&D work on solvent extraction was provided not by WSRC, but through DOE's Office of Science and Technology. Part of this funding inequity can be traced to the late 1998 decision by WSRC to pursue only one primary (small tank TPB) and one backup (CST ion exchange) option for processing the cesium-bearing salt solutions. However, this funding disparity appears to be primarily responsible for the different levels of technical maturities of the four processing options, independent of their likelihoods of success. The committee got the sense from its discussions with WSRC and DOE staff at the information-gathering meeting that WSRC did not appear to be serious about pursuing R&D on any option but small tank TPB precipitation.

Second, WSRC does not appear to have a well thought out R&D plan for the small tank TPB precipitation and CST ion exchange options[15]. There are no written R&D plans for either of these options. Moreover, in response to committee questions, the WSRC and DOE participants at the committee's information-gathering meeting were unable to describe an R&D scope for these options that would resolve the outstanding issues. Rather, the committee was presented with lists of research needs, but these needs were not prioritized. The committee was puzzled

---

[15] The committee learned at its information-gathering meeting that WSRC has discontinued all R&D work on the caustic side solvent extraction or direct grouting options.

Ernest J. Moniz
October 14, 1999
Page 8

by the lack of program planning and pursuit of important uncertainties. Even for the WSRC-favored small tank TPB processing option, R&D planning and establishment of priorities apparently have not been done.

   • Conclusions and Recommendations. As noted previously, DOE plans to release a draft EIS in October 1999 that will be used as a basis for a spring 2000 ROD that selects a single processing option. WSRC is now in the process of preparing the draft EIS, and the committee was told by WSRC staff that the draft EIS would likely recommend the selection of small tank TPB precipitation as the preferred processing option. Based on the committee's initial review of the processing options, it is not clear that the small tank TPB precipitation option favored by WSRC will necessarily be the committee's preferred choice after the committee's detailed review is completed.

   Although the committee recognizes the need for selecting and implementing an option as soon as possible—both because of the high ($400 million per year) operating costs for the high-level waste system at Savannah River and because of the potential hazard posed by the aging waste tanks—the committee concludes from the preceding discussion that there are significant technical or regulatory risks in selecting any of the four options, including the small tank TPB precipitation option that is apparently preferred by WSRC. **Therefore, the committee recommends that WSRC pursue vigorously one primary and several backup options for processing the cesium-bearing salt solutions at Savannah River until the remaining technical and regulatory issues are resolved. This may require that DOE delay the issuance of the planned EIS and ROD, or that DOE adopt a phased-decision approach in the EIS and ROD that would allow several processing options to be pursued in parallel until a clearly superior option is identified. DOE and WSRC should enlist the help of specialists in industry, National Laboratories, universities, and other federal agencies to resolve these scientific, technical, and regulatory issues.**

   To meet the ambitious time schedule for selecting and implementing any processing option, the committee concludes that DOE and WSRC will have to develop and implement a sharply focused R&D program to resolve the open issues. To this end, the committee offers the following recommendations:

   1. Actinide and Strontium Removal. WSRC should continue its efforts to address the remaining technical questions concerning reaction kinetics of the MST process for removal of actinides and strontium from the tank wastes and get on to pilot-scale testing as soon as possible.

   2. Small Tank TPB Precipitation. WSRC should develop and implement a vigorous, well-planned, and adequately funded R&D program to address the remaining scientific hurdles with the small tank TPB precipitation. R&D should, at a minimum, address TPB decomposition process(es) and the envelope of catalysts and catalytic reactions responsible for benzene generation.

   3. CST Ion Exchange. A vigorous, well planned, and adequately funded R&D effort should be undertaken to address the remaining scientific hurdles with the CST ion exchange option. This R&D should address, at a minimum, the cesium desorption process and reactions between CST and the alkaline waste. This effort also should be pursued independently of WSRC, which does not have the needed R&D expertise on site.

Ernest J. Moniz
October 14, 1999
Page 9

4. **Direct Disposal in Grout.** WSRC and DOE should undertake a vigorous program to determine the regulatory acceptability of the direct grout option through discussions with relevant staff at DOE, the U.S. Nuclear Regulatory Commission, the U.S. Environmental Protection Agency, and South Carolina Department of Health and Environmental Control. These discussions should focus on the likely feasibility of demonstrating compliance with regulations and the strategy needed to achieve regulatory approvals.

5. **Caustic Side Solvent Extraction.** A vigorous, well planned, and adequately funded R&D effort should be undertaken to address the remaining scientific and technical hurdles with the caustic side solvent extraction option. This R&D should address, at a minimum, the stability of the solvent system in radiation fields, the ability to scrub and recycle the solvents, the ability to mitigate contaminant formation during processing, and the ability to produce the chelating agent in quantities necessary for this application. If started immediately, it may be possible to complete this work by next spring, in time for the final EIS. This effort should be pursued independently of WSRC, which does not have the needed R&D expertise on site for this particular solvent extraction system.

The committee's next information-gathering meeting will be held in Augusta, Georgia on November 21-22, 1999, and a major part of that meeting will be devoted to further discussions of the R&D needed to resolve the issues identified in this report. The committee plans to ask Department and WSRC staff to report on their future R&D plans to resolve the open issues. The committee will provide a critique of these plans in its final report, which it hopes to issue by April 2000.

Sincerely yours,

Milt Levenson, Chair
Greg Choppin, Vice Chair

Ernest J. Moniz
October 14, 1999
Page 10

ATTACHMENT A
COMMITTEE ON CESIUM PROCESSING ALTERNATIVES
FOR HIGH-LEVEL WASTE AT THE SAVANNAH RIVER SITE

MILTON LEVENSON, *Chair*, Bechtel International (retired), Menlo Park, California
GREGORY CHOPPIN, *Vice-Chair,* Florida State University, Tallahassee
JOHN BERCAW, California Institute of Technology, Pasadena
DARYLE BUSCH, University of Kansas, Lawrence
TERESA FRYBERGER, Brookhaven National Laboratory, Upton, New York
GEORGE KELLER, Union Carbide Corporation (retired), South Charleston, West Virginia
MATTHEW KOZAK, Monitor Scientific, LLC, Denver, Colorado
ALFRED SATTELBERGER, Los Alamos National Laboratory, Los Alamos, New Mexico
BARRY SCHEETZ, The Pennsylvania State University, University Park
MARTIN STEINDLER, Argonne National Laboratory (retired), Downers Grove, Illinois

NRC Staff

KEVIN CROWLEY, Study Director
DOUGLAS RABER, Director, Board on Chemical Sciences and Technology
ROBERT ANDREWS, Senior Staff Officer, Board on Radioactive Waste Management
JOHN WILEY, Senior Staff Officer, Board on Radioactive Waste Management
LATRICIA BAILEY, Project Assistant
MATTHEW BAXTER-PARROTT, Project Assistant

Ernest J. Moniz
October 14, 1999
Page 11

ATTACHMENT B
LETTER OF REQUEST FOR THIS STUDY

The Under Secretary of Energy
Washington, DC 20585

June 28, 1999

Dr. Bruce Alberts
Chair, National Research Council
National Academy of Sciences ·
2101 Constitution Avenue, N.W.
Washington, D.C. 20418

Dear Dr. Alberts:

I am writing to request that the National Research Council conduct an independent
technical review of the alternatives the Department of Energy is considering for
processing the high-level radioactive waste (HLW) salt solutions at the Savannah
River Site (SRS). The Department will probably select one of these alternatives as
the preferred alternative. We will use the other alternatives as backup/technically
viable to replace the In-Tank Precipitation (ITP) process on which work was
stopped recently because of technical problems.

The SRS was established during the early 1950s in support of the defense mission
to produce plutonium and other materials for nuclear weapons. The 40-plus years
of nuclear material production at the SRS has resulted in the generation of
approximately 34 million gallons of high-level radioactive waste, which is currently
stored in large underground tanks at the site.

About 90 percent of this volume is comprised of salts and salt solutions that
contain high levels of radioactive cesium, which was to be removed by the ITP
process prior to treatment and immobilization. A systems engineering evaluation
of all high-level waste salt separation processes was recently completed as a result
of the problem with the ITP process. The Department is now considering three
alternative processes to address the cesium removal problem.

The treatment of the HLW salts is a complex and costly technical challenge for the
Department. The Department has spent about $489 million since 1983 at the SRS
to design and construct facilities for cesium removal and is faced with the
challenge of selecting and implementing an alternative process. We are proceeding
with some urgency to identify an alternative to avoid costly disruptions to ongoing
waste processing activities at the site. I am committed to ensuring that our
decisions on a path forward have a sound technical basis.

Ernest J. Moniz
October 14, 1999
Page 12

2

For several decades, the National Research Council's advice to the Department has been helpful in its efforts to bring good science and technology to bear in the environmental management program. I believe that the Council can now assist the Department as we proceed to identify and implement an alternative process for processing the HLW salt solutions at the SRS. Therefore, I would like to request that the Council review and make recommendations on the alternative options that have been recommended for processing the HLW salt solutions at the SRS. I would like this review to address the following points:

- Was an appropriately comprehensive set of cesium partitioning alternatives identified and are there other alternatives that should be explored?

- Was the process used to screen the alternatives technically sound and did its application result in the selection of appropriate preferred alternatives?

- Are there significant barriers to the implementation of any of the preferred alternatives, taking into account their state of development and their ability to be integrated into the existing SRS HLW system?

- Are the planned R&D activities, including pilot-scale testing, adequate to support implementation of a single preferred alternative?

Members of my staff have been in contact with Dr. Kevin Crowley of the Board on Radioactive Waste Management to discuss this project and develop the work scope for a National Research Council review. I would like this review to begin immediately so that we can use the Council's advice prior to making a final decision in April 2000.

Additionally, I would like to receive a preliminary report from the Council by the end of September 1999 that identifies significant issues or problems with the alternatives so that we can factor this advice into our draft environmental impact statement (EIS). Following issuance of the draft EIS, I would expect the availability of your final report during the public comment period.

Ernest J. Moniz
October 14, 1999
Page 13

3

Mr. Ralph Erickson and his staff are the Department's principal points of contact
for this work. The Environmental Management Program will fund this project
under the cooperative agreement with the Council's Board on Radioactive Waste
Management (DE-FC01-99EW59049).

I appreciate the Council's help on this very important project.

Sincerely,

Ernest J. Moniz

Ernest J. Moniz
October 14, 1999
Page 14

ATTACHMENT C
BACKGROUND ON THE HIGH-LEVEL WASTE PROGRAM AT SAVANNAH RIVER

During and immediately following the Second World War, the U.S. Government established large industrial complexes at several sites across the United States to develop, manufacture, and test nuclear weapons. One of these complexes was established in 1950 at the Savannah River Site to produce strategic isotopes, mainly plutonium and tritium, for defense purposes. The site is located adjacent to the Savannah River near the Georgia-South Carolina border (Figure C.1) and comprises an area of about 800 square kilometers (~300 square miles).

The Savannah River Site is host to an extensive complex of facilities that includes fuel and target fabrication plants, nuclear reactors, chemical processing plants, underground storage tanks, and waste processing and immobilization facilities. Plutonium and tritium were produced by irradiating specially prepared metal targets in the nuclear reactors at the site. After irradiation, the targets were transferred to the *F Canyon* or *H Canyon*, where they were processed chemically to recover these isotopes. This processing resulted in the production of large amounts of highly radioactive liquid waste, known as *high-level waste* (HLW), that is being stored in two underground tank farms at the site (in the *F Tank Farm* and *H Tank Farm*). The U.S. Department of Energy (DOE) has the responsibility for waste management at the Savannah River Site and has implemented a program to stabilize this high-level waste and close the tank farms.

The information used in this attachment is taken from the documents cited in Attachment E and from copies of the presentations provided at the committee's first information-gathering meeting.

HIGH-LEVEL WASTE SYSTEM AT SAVANNAH RIVER SITE

A simplified schematic representation of the HLW system at the Savannah River Site is shown in Figure C.2. This system comprises the following major components:

Waste Concentration and Storage. The high-level waste resulting from operations in the F and H Canyons is currently being stored in 48 underground carbon-steel tanks[16] in the F and H Tank Farms. The tanks range in size from about three million to five million liters (750,000 to 1.3 million gallons). The high-level waste was made alkaline with NaOH before being transferred to the tanks to reduce corrosion of the carbon steel primary containment. Consequently, the waste has a high pH (>14) and a high salt content.

Approximately 400 million liters (100 million gallons) of high-level waste have been produced at Savannah River since operations began in the 1950s, but this volume has been reduced to about 130 million liters (34 million gallons) by removal of excess water through evaporator processing operations. About 10% of the waste by volume is in the form of a metal precipitate, or *sludge*, that contains most of the actinides and fission products. This sludge was formed by natural settling and by precipitation when NaOH was added to the waste. The remaining waste consists of salt in a solid, or *saltcake*, form that contains cesium and minor amounts of actinides and other fission products. Salt was produced by processing the alkaline

---

[16] There are 51 tanks in the F and H Tank Farms, but only 48 contain waste at present. Two tanks have been filled with grout and one tank is empty.

Ernest J. Moniz
October 14, 1999
Page 15

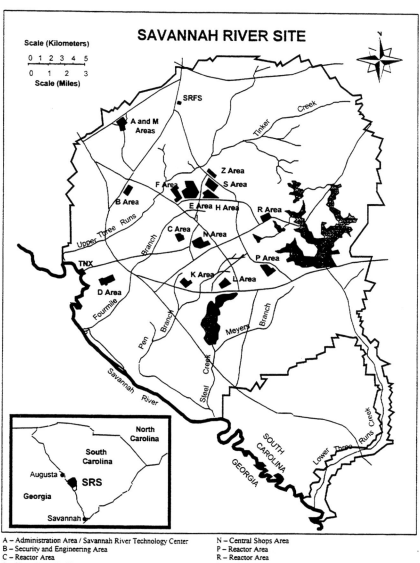

SAVANNAH RIVER SITE

Scale (Kilometers)

0 1 2 3 4 5

0 1 2 3

Scale (Miles)

SRFS

A and M Areas

Tinker Creek

Z Area

F Area   S Area

B Area

E Area   H Area   R Area

Upper Three Runs

Branch

C Area   N Area

TNX

P Area

K Area   L Area

D Area

Fourmile

Pen Branch

Meyers Branch

Savannah River

Steel Creek

Lower Three Runs Creek

North Carolina

South Carolina

Augusta

SRS

Georgia

SOUTH CAROLINA

GEORGIA

Savannah

A – Administration Area / Savannah River Technology Center
B – Security and Engineering Area
C – Reactor Area
D – Heavy Water Facility and Power House
E – Burial Grounds
F – Separations Area / F-Tank Farm
H – Separations Area / H-Tank Farm
K – Reactor Area
L – Reactor Area
M – Materials Area

N – Central Shops Area
P – Reactor Area
R – Reactor Area
S – Defense Waste Processing Facility
T – Large Scale Test Facility
Z – Saltstone Area
RR – RailRoad Yard
SR Forest Station
PAR Pond – Manmade cooling pond for P and R reactors
L Lake – Manmade cooling pond for L reactors

FIGURE C.1. Map showing location of the Savannah River Site

Ernest J. Moniz
October 14. 1999
Page 16

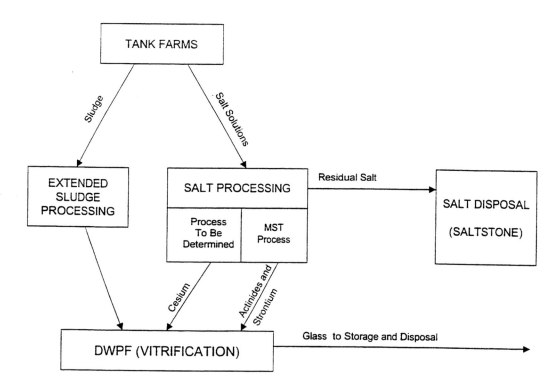

FIGURE C.2. High Level Waste System at the Savannah River Site

Ernest J. Moniz
October 14, 1999
Page 17

waste through evaporators to remove excess water. After processing, the waste was pumped back into the tanks, where it cooled and crystallized.

Radionuclide Immobilization. The Defense Waste Processing Facility (DWPF) was constructed to immobilize radioactive waste in borosilicate glass for eventual shipment to and disposal in a geological repository. The glass-making process is referred to as *vitrification.* This glass is produced by combining the processed high-level waste (the processing operations are discussed below) with specially formulated glass frit and melting the mixture at about 1150 °C. The molten glass is then poured into cylindrical stainless steel canisters, allowed to cool, and sealed. The DWPF canisters are about 60 centimeters (2 feet) in diameter and about 300 centimeters (10 feet) in length and contain about 1,800 kilograms (4,000 pounds) of glass. About 700 canisters have been produced to date, and Savannah River estimates that a total of about 6,000 canisters will be produced by 2026, when the tank waste processing program is planned to be completed. These canisters will be stored at the site until a repository is opened and ready to receive them.

Extended Sludge Processing is used to prepare the sludge portion of the tank waste for processing into glass. The sludge is removed from the tanks by hydraulic slurrying, and it is then washed to remove aluminum and soluble salts, both of which can interfere with the glass-making process. The washed sludge is transferred to the DWPF for further processing before being incorporated into glass.

Salt Processing will be used to remove radionuclides from the HLW salt for eventual processing into glass. The salt will be redissolved and transferred out of the tanks. It will then be mixed with a sorbent to remove any remaining actinides (mainly uranium and plutonium) and strontium. The currently planned sorbent is monosodium titanate (MST). The solution will then be subjected to another (and as-yet undetermined) process to remove cesium. This processing step is the focus of the present study. The separated actinides, strontium, and cesium will be washed to remove soluble salts and sent to the DWPF for immobilization.

Salt Disposal. A variety of secondary waste streams are formed during the processing operations described above. Some of these waste streams are recycled back to the tanks, other wastes are recycled within the various processing operations, and yet other wastes are treated and released to the environment. Most notably, the residual salt solutions (i.e., the solutions remaining after actinide and cesium removal) will be disposed of onsite in a waste form known as *saltstone.* The residual solutions are classified as "incidental waste" from the processing of high-level waste. Saltstone is created by mixing the residual salt solutions with fly ash, slag, and Portland cement to create a grout slurry. This slurry is then poured into concrete vaults, where it cures and is eventually covered with soil. The saltstone also contains some radionuclides, for example technetium-99 and tin-126. The Saltstone Production Facility is permitted by the South Carolina Department of Health and Environmental Control as waste water treatment facility. The saltstone vaults are designed as a controlled release landfill disposal site. The operating permit limits the average concentrations of radioactive contaminants to the limits specified by the U.S. Nuclear Regulatory Commission for Class A Waste[17]. In the direct disposal in grout processing option, which is discussed below, cesium also would be immobilized in the grout.

At present, Savannah River is processing sludge from the tanks to make glass at the DWPF, and it has a wastewater permit from South Carolina to produce saltstone. The current

---

[17] As provided in 10 CFR Part 61.

Ernest J. Moniz
October 14, 1999
Page 18

high-level waste processing schedule calls for the salt solutions to be processed to recover the actinides, strontium, and cesium beginning about 2008. The 2008 schedule has been proposed to maintain operations at the DWPF and to ensure that there is sufficient space in the tank farms to continue operations at the site[18]. To meet this schedule, however, Savannah River must develop, test, and implement a process for removing actinides, strontium, and cesium from the salt in the tanks. A brief review of Savannah River's efforts to develop this process is provided in the next section.

SALT PROCESSING OPTIONS

The objective of the salt processing step is to reduce the volume of salt waste to be immobilized in glass and, consequently, to reduce the time and cost of the immobilization operations. There are approximately 120 million liters (31 million gallons) of HLW salt in the F and H Tank Farms, but Savannah River estimates that this salt could be processed to yield about 11 million liters (3 million gallons) of actinide- and cesium-bearing solutions or precipitates for vitrification, roughly a ten-fold reduction in volume.

At present, Savannah River plans to remove actinides, strontium, and cesium from the salt solutions in two processing steps. As noted previously, actinides and strontium will be removed by mixing the salt solutions with MST. The resulting reaction leads to the sorption of actinides and strontium. The product of this reaction could be removed from the salt solutions by filtration for subsequent processing and immobilization. This process has been demonstrated in conjunction with the in tank precipitation program, which is discussed below, but additional R&D is under way to resolve some remaining problems.

The removal of cesium from the salt solutions is potentially feasible through a number of processes, for example, precipitation reactions, ion exchange, or solvent extraction. In the 1980's, Savannah River developed a process for removing cesium from salt solutions through a precipitation reaction involving sodium tetraphenylborate (TPB):

$$NaTPB + Cs^+ \Leftrightarrow Na^+ + TPB^- + Cs^+ \Leftrightarrow CsTPB + Na^+$$

The Savannah River Site refers to this process as in-tank precipitation (ITP). The TPB was to be added directly to a large waste tank to produce a cesium-bearing precipitate, which could then be processed like tank sludge. Savannah River undertook an ITP pilot project in 1983 to demonstrate proof of principle. The process removed cesium from the salt solution, but it also resulted in the generation of benzene from radiolytic reactions and possibly from catalytic reactions with trace metals in the waste, in particular, palladium and copper.

In September 1995, Savannah River initiated ITP processing operations in a tank that contained about 1.7 million liters (450,000 gallons) of salt solutions. The operations were halted after about three months because of higher-than-expected rates of benzene generation. Savannah River staff then initiated a research program to understand the mechanisms of benzene generation and release, and staff also considered possible design changes so that the benzene, which is highly flammable, could be handled safely during processing operations. In 1996, the Defense Nuclear Facility Safety Board (DFNSB) issued Recommendation 96-1, which urged DOE to halt all further testing and to begin an investigative effort to understand the mechanisms of benzene formation and release (DNFSB, 1996):

---

[18] At present, the F and H Tank Farms have about 2.6 million liters (700,000 gallons) of empty space.

Ernest J. Moniz
October 14, 1999
Page 19

> *The additional investigative effort should include further work to (a) uncover the reason for the apparent decomposition of precipitated TPB in the anomalous experiment, (b) identify the important catalysts that will be encountered in the course of ITP, and develop quantitative understanding of the action of these catalysts, (c) establish, convincingly, the chemical and physical mechanisms that determine how and to what extent benzene is retained in the waste slurry, why it is released during mixing pump operation, and any additional mechanisms that might lead to rapid release of benzene, and (d) affirm the adequacy of existing safety measures or devise such additions as may be needed.*

Investigations by Savannah River in 1997 uncovered the possible role of metal catalysts in the benzene formation process. However, Savannah River concluded that both safety and production requirements could not be met, which led to the suspension of operations altogether in early 1998. At the time of suspension, Savannah River had spent $489 million to develop and implement the ITP process.

In March 1998, the Savannah River contractor, Westinghouse Savannah River Company (WSRC), formed a systems engineering team to identify alternatives to the ITP process for separating cesium. This team was comprised of 10 members with expertise in science and engineering, operations, waste processing, and safety and regulations. The team interacted with experts throughout the DOE complex and undertook a historical review and literature survey to identify about 140 processes that could potentially be used to separate cesium from the salt solutions. These processes were grouped into an "initial list" of 18 alternative processing options, which were subsequently screened using a multi-attribute analysis to obtain a "short list" of four alternative processing options: (1) small tank tetraphenylborate (TPB) precipitation, (2) crystalline silicotitanate (CST) ion exchange, (3) caustic side solvent extraction, and (4) direct disposal in grout.

Small tank TPB precipitation is carried out in specially designed processing vessels to control benzene generation. TPB is the same precipitating agent used in the ITP process. The process allows for closer temperature control and faster cycling times to reduce the generation of benzene and improved agitation of the liquid to facilitate benzene removal. The process is also designed with secondary containment and positive pressure control so that the processing vessels could be blanketed with nitrogen to reduce explosion hazards and facilitate benzene removal. The process generates a precipitate slurry that will be transferred to the DWPF.

CST ion exchange is based on conventional ion exchange concepts but utilizes a non-elutable inorganic solid that has a high selectivity for cesium over other alkali metals. The waste would be processed by pumping it through columns packed with this material. As the salt solutions pass through the CST, cesium is trapped. Once loaded with cesium, the CST would then be sent directly to the DWPF for further treatment and vitrification. Although ion exchange for cesium removal has been used in the nuclear industry, CST ion exchange has never been used in a large-scale waste application, and CST has never been manufactured in commercial-scale quantities.

Caustic side solvent extraction also is based on conventional solvent extraction concepts, such as those used in the widely known PUREX process to separate U and Pu from dissolved irradiated targets. The process involves the mixing and subsequent separation of two immiscible feed streams: an aqueous solution containing the radionuclide to be extracted and

Ernest J. Moniz
October 14, 1999
Page 20

an organic solvent containing a chelating agent (also known as the *extractant*) for that radionuclide. Other chemicals may be added to improve the extraction efficiency or to inhibit the formation of undesirable reaction products.

The two feed streams are pumped through a series of centrifugal contactors, where they are mixed and subsequently separated on the basis of density. During the mixing process, the radionuclide is chelated by the extractant, which results in its transfer from the aqueous feed stream to the organic feed stream. The radionuclide is then recovered through a series of stripping steps, and the organic solvent is recycled back into the front end of the extraction process.

For caustic side solvent extraction of cesium, the organic solvent consists of a *diluent* (Isopar®-L, a mixture of branched alkanes), *modifier* (Cs-3, a fluorinated alcohol that prevents the formation of additional chemical phases), and *extractant* (BOB Calix [19], a calixarene crown ether). When the salt solution is mixed with the organic solvent, cesium ions are complexed by the extractant ("L" in the following reaction) to form a cesium nitrate ion pair:

$$Cs^+ + NO_3^- + L \Leftrightarrow [CsL]^+ NO_3^-$$

This ion pair is subsequently extracted into the organic solvent and then is recovered by separation of the aqueous stream from the solvent stream followed by a series of acid washing steps. The Isopar®-L and BOB Calix are recycled, and the cesium nitrate liquid can be sent directly to the DWPF without further processing.

Although solvent extraction is a mature technology for separating radionuclides from acid solutions (acid side solvent extraction), solvent extraction of cesium from highly alkaline solutions has never been demonstrated on an industrial scale, and the chelator (BOB Calix) has never been produced in commercial quantities. It is currently being manufactured in small quantities and is priced at about $500 per gram. The price could presumably be reduced significantly once production was scaled up.

Direct disposal in grout is very similar to the saltstone process that was to have been used to immobilize the residual salt solutions from ITP operations. After removal of the actinides and strontium with MST, the cesium-bearing salt solutions would be mixed with fly ash, slag, and Portland Cement and poured into concrete vaults on the site. About 520,000 cubic meters (18 million cubic feet) of grout would be produced. This waste would meet the limits of Class C low-level waste[20].

Savannah River performed a flowsheet analysis, risk analysis, and lifecycle cost analysis of the four alternatives in late 1998. Based on this analysis, WSRC recommended small tank TPB as the preferred alternative and CST ion exchange as the backup alternative. However, DOE's analysis of the information concluded that additional research was needed before a preferred alternative could be selected. Additional R&D needs were identified for each of these alternatives, and work to address these needs was underway as the committee began this study.

---

[19] Formally, Calix[4]arene-bis(t-octylbenzo-crown-6).

[20] The total cesium-137 activity of the grout would be about 120 million curies.

Ernest J. Moniz
October 14, 1999
Page 21

ATTACHMENT D
LIST OF REPORT REVIEWERS

This letter report has been reviewed in draft form by individuals chosen for their diverse perspectives and technical expertise, in accordance with procedures approved by the NRC's Report Review Committee. The purpose of this independent review is to provide candid and critical comments that will assist the institution in making the published report as sound as possible and to ensure that the report meets institutional standards for objectivity, evidence, and responsiveness to the study charge. The review comments and draft manuscript remain confidential to protect the integrity of the deliberative process. We wish to thank the following individuals for their participation in the review of this report:

Andrew Campbell, U.S. Nuclear Regulatory Commission
Rodney Ewing, University of Michigan
Mary Good, Venture Capital Investors, LLC
Michael Kavanaugh, Malcolm-Pirnie, Inc.
Tobin Marks, Northwestern University
Royce Murray, University of North Carolina
Kenneth Raymond, University of California, Berkeley
Robin Rogers, University of Alabama
Vincent Van Brunt, University of South Carolina

While the individuals listed above have provided constructive comments and suggestions, it must be emphasized that responsibility for the final content of this report rests entirely with the authoring committee and the institution.

Ernest J. Moniz
October 14, 1999
Page 22

ATTACHMENT E
DOCUMENTS RECEIVED BY THE COMMITTEE

Case, Joel. 1998. Memorandum to James Owendoff Regarding the Savannah River High Level Waste Salt Disposition Independent Project Evaluation Team Review And Assessment (OPE-HLW-99-005). December 26.

Defense Nuclear Facilities Safety Board. 1996. Recommendation 96-1 to the Secretary of Energy pursuant to 42 U.S.C. 2286(a) (5) Atomic Energy Act of 1954, as amended. August 14.

Department of Energy. 1998. Savannah River Review Team Final Report on the High Level Waste Salt Disposition Alternatives Evaluation. December.

General Accounting Office. 1999. Nuclear Waste: Process to Remove Radioactive Waste from Savannah River Tanks Fails to Work. GAO/RCED-99-69. April.

Independent Evaluation Team. 1998. Independent Assessment of the Savannah River Site High-Level Waste Salt Disposition Alternatives Evaluation. DOE/ID-10672. December.

McCullough, J.W. 1999. Department of Energy, Savannah River Management Plan for Phase IV of the High Level Waste Salt Disposition Alternatives Evaluation. April.

McCullough, J.W., and P.C. Suggs. 1998. Department of Energy, Savannah River Review Team Report on the HLW Salt Disposition Alternatives. July 28.

Papouchado, L., E. Kosiancic, J. Carlson, and P. Suggs. 1998. Trip Report—Site Visited: Hanford Site visited May 7 and 8. May 22.

Papouchado, L., E. Kosiancic, J. Carlson, and P. Suggs. 1998. Trip Report—Sites Visited: Willowcreek (INEEL), ICPP (INEEL). May 11.

Perella, V. 1998. Memorandum to Steve Piccolo Regarding the Results of the HLW Salt Disposition Systems Engineering Team, Phase II: Criteria Selection and Weighting For HLW Salt Disposition "Initial List" Down Selection. June 9.

Piccolo, Steve. 1998. Candidate Selections for the HLW Salt Disposition Systems Engineering Team. Revision 0. March 25.

Piccolo, Steve. 1998. Candidate Selections for the HLW Salt Disposition Systems Engineering Team. Revision 1. September 15.

Piccolo, S., K. Reuter, P. Hudson, J. Barnes, and B. Spader. 1998. Trip Report—Site Visited: Sellafield. June 2.

Poirier, M.R. 1998. Memorandum to Steve Piccolo Regarding the Evaluation of Potential Cesium Removal Technologies. SRT-WPT-98-008. June 5.

Ernest J. Moniz
October 14, 1999
Page 23

Poirier, M.R., R. D. Hunt, and C. Carlson. 1998. Identification of Cesium Removal Technologies. WSRC-TR-98-00181. May 29.

Rudy, Greg. 1998. Memorandum to James Owendoff Regarding Program Plan for the Evaluation of High Level Waste Salt Disposition Alternatives. March 16.

Rueter, K., P. Hudson, E. Murphy, and P. Suggs. 1998. Trip Report—Site Visited: Oak Ridge National Laboratory on May 21, 1998. May 31.

Rueter, K., E. Murphy, and P. Suggs. 1998. Trip Report—Site Visited: West Valley Site on May 19. June 1.

Savannah River Site, High Level Waste Salt Disposition Team. 1998. Identification of Alternatives Briefing Package. March 12.

Savannah River Site, High Level Waste Salt Disposition Team. 1998. Position Paper on the Evaluation Leading to the "Initial List" of Alternatives. April 17.

Savannah River Site, High Level Waste Salt Disposition Team. 1999. HLW Salt Disposition Alternatives Identification Preconceptual Phase II Summary Report. Revision 2. WSRC-RP-98-00165. June 24.

Savannah River Site, High Level Waste Salt Disposition Systems Engineering Team. 1998. Position Paper on Identifying Alternatives to the In-Tank Precipitation Process. March 24.

Savannah River Site, High Level Waste Salt Disposition Systems Engineering Team. 1998. Preconceptual, Phase I Initial Design Input. April 2.

Savannah River Site, High Level Waste Salt Disposition Systems Engineering Team. 1998. Systems Engineering Management Plan for Development of Alternatives to Process and Dispose of High Level Waste Salt. WSRC-RP-98-00163. Revision 0. April 17.

Savannah River Site, High Level Waste Salt Disposition Systems Engineering Team. 1998. HLW Salt Disposition Alternatives Identification Preconceptual Phase I, Summary Report. WSRC-RP-98-00162. April 17.

Savannah River Site, High Level Waste Salt Disposition Systems Engineering Team. 1998. Position Paper on Identifying and Documenting Dissenting Opinions in the Evaluation and Selection of Alternatives to the In-Tank Precipitation Process. HLW-SDT-980005. May 15.

Savannah River Site, High Level Waste Salt Disposition Systems Engineering Team. 1998. Position Paper on the Approach to Flowsheet Analysis. Revision 0. HLW-SDT-980009. May 29.

Savannah River Site, High Level Waste Salt Disposition Systems Engineering Team. 1998. Position Paper on the Approach to Information Handling, Analysis and Reporting. Revision 0. HLW-SDT-980010. June 4.

Ernest J. Moniz
October 14, 1999
Page 24

Savannah River Site, High Level Waste Salt Disposition Systems Engineering Team. 1998. Position Paper on Determination of Risk and Risk Handling Strategies for the Initial List Alternatives. Revision 1. HLW-SDT-980004. June 4.

Savannah River Site, High Level Waste Salt Disposition Systems Engineering Team. 1998. Position Paper on Dispositioning of Pro-Formas Received During Phase II. Revision 0. HLW-SDT-980014. June 8.

Savannah River Site, High Level Waste Salt Disposition Systems Engineering Team. 1998. Position Paper on Preliminary Life Cycle Cost Analysis to Select the Short List of Alternatives. Revision 0. HLW-SDT-980013. June 8.

Savannah River Site, High Level Waste Salt Disposition Systems Engineering Team. 1998. Pre-conceptual, Phase I Initial Design Input. June 24 Revision.

Savannah River Site, High Level Waste Salt Disposition Systems Engineering Team. 1998. Results Report on Preliminary Life Cycle Cost Estimates for Initial List Alternatives. Revision 0. HLW-SDT-980018. June 25.

Savannah River Site, High Level Waste Salt Disposition Systems Engineering Team. 1998. Position Paper on the Sensitivity Analyses of Alternative Methods for Disposition of High Level Salt Waste. Revision 0. WSRC-TR-98-00236. June 26.

Savannah River Site, High Level Waste Salt Disposition Systems Engineering Team. 1998. Systems Engineering Management Plan for Development of Alternatives to Process and Dispose of High Level Waste Salt. Revision 1. WSRC-RP-98-00163. August 21.

Savannah River Site, High Level Waste Salt Disposition Systems Engineering Team. 1998. Position Paper on the Use of Weighted Evaluation Criteria to Select the Short List of Alternatives. Revision 2. HLW-SDT-980006. September 17.

Savannah River Site, High Level Waste Salt Disposition Systems Engineering Team. 1998. Results Report on Preliminary Risk Assessment with Adjusted Risk Values. Revision 1. HLW-SDT-980015. October 14.

Savannah River Site, High Level Waste Salt Disposition Systems Engineering Team. 1998. Results Report on Utility Function Evaluation. Revision 1. HLW-SDT-980019. October 14.

Savannah River Site, High Level Waste Salt Disposition Systems Engineering Team. 1998. Final Report (3 Volumes), Recommendation Preconceptual Design and Initial Cost Estimate. WSRC-RP-98-00170. October 29.

Savannah River Site, High Level Waste Salt Disposition Systems Engineering Team. 1999. Applied Technology Integration Scope of Work Matrix for Decision Making (Small Tank TPB Precipitation, CST Non-Elutable Ion Exchange and Direct Disposal of Grout) HLW-SDT-99-0009. April 14.

Scott, A.B. 1990. Letter to Roy Schepens Regarding the HLW Salt Disposition Systems Engineering Team Charter and Attached Charter. March 13.

Ernest J. Moniz
October 14, 1999
Page 25

Westinghouse Savannah River Company. 1998. High Level Waste Salt Disposition Interface Requirements. WSRC-RP-98-00164. May 20.

Westinghouse Savannah River Company. 1998. Bases, Assumptions, and Results of the Flowsheet Calculations for the Initial Eighteen Salt Disposition Alternatives. Revision 1. WSRC-RP-98-00166. September 15.

Westinghouse Savannah River Company. 1999. Briefing Packages for the National Research Council Committee on Cesium Processing Alternatives for High-Level Waste at the Savannah River Site. Presentations given September 13 and 14. Unpublished.

Westinghouse Savannah River Company. Undated. Draft High Level Waste Salt Disposition Interface Requirements. Revision C.

Ernest J. Moniz
October 14, 1999
Page 26

ATTACHMENT F
PRESENTATIONS GIVEN DURING FIRST COMMITTEE MEETING

Background on Cesium Separations at Savannah River.
Steve Piccolo, Westinghouse Savannah River Company (WSRC)
Roy Schepens, U.S. Department of Energy

Background on the In-Tank Precipitation (ITP) Process.
Joe Carter, WSRC
Walt Tamosaitis, Savannah River Technology Center (SRTC)
Mark Barnes, SRTC
Roy Jacobs, WSRC
Mike Montini, WSMS

Small Tank TPB Option.
Sam Fink, SRTC
Reid Peterson, SRTC
Mark Barnes, SRTC
Hank Elder, WSRC
Jack Collins, Oak Ridge National Laboratory (ORNL)
David Hobbs, SRTC

CST Non-Elutable Ion Exchange Option.
Doug Walker, SRTC
Roy Jacobs, WSRC
John Harbour, SRTC
Dan Lambert, SRTC
Bill Wilmarth, SRTC

Caustic Side Solvent Extraction Option.
Ken Rueter, WSRC
Ralph Leonard, Argonne National Laboratory
Bruce Moyer, ORNL
Reid Peterson, SRTC
John Fowler, WSRC

Direct Disposal in Grout Option.
Ed Stevens, SRTC
Christine Langton, SRTC
Jim Cook, SRTC
Elmer Wilhite, SRTC

Ernest J. Moniz
October 14, 1999
Page 27

ATTACHMENT G
STATEMENT OF TASK

The committee will review the Department of Energy's work to identify alternatives for separating cesium from high-level waste at the Savannah River site. This review will address the following points:

- Was an appropriately comprehensive set of cesium partitioning alternatives identified, and are there other alternatives that should be explored?
- Was the process used to screen the alternatives technically sound and did its application result in the selection of appropriate preferred alternatives?
- Are there significant barriers to the implementation of any of the preferred alternatives, taking into account their state of development and their ability to be integrated into the existing SRS HLW system?
- Are the planned R&D activities, including pilot-scale testing, adequate to support implementation of a single preferred alternative?

Appendix C

Information-Gathering Meetings

PRESENTATIONS GIVEN DURING FIRST COMMITTEE MEETING
September 13–14, 1999, Augusta, GA

Background (Steve Piccolo and Roy Schepens)

Large Tank ITP(Joe Carter, WSRC; Walt Tamosaitis, SRTC; Mark Barnes, SRTC; Roy Jacobs, WSRC-HLW; Mike Montini, WSMS)

Small Tank TPB presentation (Sam Fink, SRTC; Reid Peterson, SRTC; Mark Barnes, SRTC; Hank Elder, WSRC-HLW; Jack Collins, ORNL; David Hobbs, SRTC)

CST Non-Elutable ion exchange presentation (Doug Walker, SRTC; Roy Jacobs, WSRC-HLW; John Harbour, SRTC; Dan Lambert, SRTC; Bill Wilmarth, SRTC)

Caustic Side Solvent Extraction presentation (Ken Rueter, WSRC-HLW; Ralph Leonard, ANL; Bruce Moyer, ORNL; Reid Peterson, SRTC; John Fowler, WSRC-HLW)

Direct Disposal in Grout presentation (Ed Stevens, SRTC; Christine Langton, SRTC; Jim Cook, SRTC; Elmer Wilhite, SRTC)

PRESENTATIONS GIVEN DURING SECOND COMMITTEE MEETING
November 21–22, 1999, Augusta, GA

Discussion of the Committee's Interim Report (Frank McCoy, DOE; Susan Wood, DOE; Karen Patterson, Citizens Advisory Board)

Screening Process and Alternatives Selection, Including Discussion of Reasons for Exclusion of Processes Used at Hanford and West Valley (Steve Piccolo, Chair, Screening Committee)

TPB Precipitation—Scientific, Technical, and Regulatory Issues; R&D Plans (Sam Fink and Mark Barnes, SRTC)

Ion Exchange (CST, other)—Scientific, Technical, and Regulatory Issues; R&D Plans (Bill Wilmarth, SRTC; Dan McCabe, SRTC; John Sherman, UOP)

Caustic Side Solvent Extraction—Scientific, Technical, and Regulatory Issues; R&D Plans (Peter Bonneson, ORNL)

"Front-End" Actinide and Strontium Removal (MST) Scientific, Technical, and Regulatory Issues; R&D Plans (David Hobbs and Sam Fink, SRTC)

Direct Grout Option—Scientific, Technical, Regulatory, and Non-Proliferation Issues; R&D Plans (John Reynolds, DOE-SR)

General Overview of Research and Development Planning (Lab and University Partnering, Integration) (Lou Papouchado and Susan Wood, SRTC)

DOE Decision Process and Path Forward and Means of Implementation (Frank McCoy)

Appendix D

Incidental Waste

Waste resulting from reprocessing spent nuclear fuel that is determined to be *incidental to reprocessing* is reclassified from being high-level waste, and is managed as transuranic or low-level waste, depending on the characteristics of the waste. In U.S. Department of Energy (DOE) Order G 435.1, two procedures were established for determining whether waste may be treated as incidental to reprocessing. DOE emphasizes that incidental waste is not a new category of waste requiring separate management, but rather a categorization of waste in accordance with its hazardous characteristics rather than its origin.

The two procedures for determining if a waste may be considered incidental to reprocessing are *citation* and *evaluation*.

- *Citation*—Waste incidental to reprocessing by citation includes spent nuclear fuel reprocessing plant wastes that meet the description included in the Notice of Proposed Rulemaking [34 FR (Federal Register) 8712] for proposed Appendix D, 10 CFR (Code of Federal Regulations) Part 50, Paragraphs 6 and 7.
- *Evaluation*—The evaluation procedure is an approach to manage waste according to its hazard, even if it was not discussed in 34 FR 8712. The evaluation procedure may lead to management of the waste as either low-level waste or as transuranic waste.

Any determination that the high-level waste supernates are incidental would lead to them being managed as low-level waste. Consequently, only the low-level waste provisions of the determination are discussed further here, and transuranic waste criteria are omitted. Incidental wastes that will be managed as low-level waste must meet three criteria to be classified as low-level waste (U.S. Nuclear Regulatory Commission, 1999):

Criterion 1: The waste must receive processing to remove key radionuclides to the maximum extent that is technically and economically practical.

Criterion 2: The waste must be shown to be managed to meet safety requirements comparable to the performance objectives set out in 10 CFR Part 61, Subpart C. Safety requirements contained in DOE Order M 435.1 Section IV are held to be comparable to those in 10 CFR Part 61 (U.S. Nuclear Regulatory Commission, 1982).

Criterion 3: DOE has established a requirement that the waste must be incorporated in a solid physical form at concentrations that do not exceed the concentration limits for Class C commercially generated low-level waste. Class C limits were established by the U.S. Nuclear Regulatory Commission (USNRC) in 10 CFR 61.55 as an upper limit for wastes generally acceptable for near-surface disposal. Alternatively, DOE may establish alternative requirements for waste classification and characterization on a case-by-case basis (DOE Order G 435.1).

ROLE OF THE U.S. NUCLEAR REGULATORY COMMISSION IN INCIDENTAL WASTE DETERMINATIONS

The USNRC expects DOE to consult with it for waste streams for which there is some question of whether the waste is high-level waste. However, the USNRC has agreed that DOE has the discretion to make incidental waste determinations (memo for Commissioner Curtiss from J.M. Taylor, January 14, 1993). Owing to the ambiguity of these positions, DOE has agreed to keep the USNRC informed of any incidental waste determinations. DOE Order G 435.1-1 has recommended that the USNRC should be consulted for all determinations using evaluation. Owing to the difficulty and associated expense of evaluation determinations, DOE has urged its sites to de-emphasize the use of this approach, and to manage high-level wastes in a manner intended to permit disposal in a geological repository.

An agreement has been reached between DOE and the USNRC that DOE will manage wastes incidental to reprocessing. This includes internal DOE performance of all regulatory functions, as is typically done for DOE-generated low-level wastes. However, DOE has agreed to consult with the USNRC on any questionable determinations that a waste is incidental to reprocessing by the evaluation procedure (DOE Order G435.1-1, page II-16).

DETERMINATION OF HIGH-LEVEL WASTE SUPERNATES AS INCIDENTAL TO REPROCESSING

Determination of high-level waste supernates as incidental to reprocessing is a key administrative and regulatory precursor step to the direct grout option. The Savannah River Site has requested a review by the USNRC of their process to categorize the residual waste in the tanks as incidental (U.S. Nuclear Regulatory Commission, 1999), but has not yet requested a review of the process to categorize waste destined for the saltstone facility. Such a categorization is needed for either the direct grout option, or to categorize residual contamination in liquid wastes following any of the separation processes under consideration.

Recently, a USNRC staff requirements memorandum was sent to DOE concerning advice from USNRC regarding the determination of incidental waste at SRS and any advice warranted on the option of direct disposal of cesium in the saltstone (U.S. Nuclear Regulatory Commission, 2000). This memorandum indicates that DOE is responsible for determining whether the waste is incidental, and that the USNRC will not propose special criteria for alternate classification of the waste as Class C Low-Level Waste.

REFERENCES CITED

U.S. Nuclear Regulatory Commission. 1982. 10 CFR Part 61–Licensing Requirements for Land Disposal of Radioactive Waste, Final Rule. Federal Register 50, 38066, Washington, D.C.

U.S. Nuclear Regulatory Commission. 1999 (December 15). Classification of Savannah River Residual Tank Waste as Incidental. SECY-99-284, Washington, D.C.

U.S. Nuclear Regulatory Commission. 2000 (May). Staff Requirements–SECY-99-0284. Classification of Savannah River Residual Tank Waste as Incidental. Memorandum from A.L.V. Cook to W.D. Travers, Washington, D.C.

Appendix E

Long-Term Safety of the Direct Grout Option

Cook (1998) conducted an initial scoping performance assessment to compare the long-term safety of the direct grout option to the safety of the existing Z-Area saltstone disposal facility, as expressed in the performance assessment for the facility (Martin Marietta Energy Systems, Inc., et al., 1992; Westinghouse Savannah River Company, 1998a). The assessment was not done using the full performance assessment methodology developed by Westinghouse Savannah River Company (Martin Marietta Energy Systems, Inc., et al., 1992; Westinghouse Savannah River Company, 1998a). Instead, the assessment comprised the following three analyses:

- a screening-level assessment[1] was conducted of doses from the groundwater pathway,
  - a simplified groundwater dose analysis was undertaken, and
  - an analysis was conducted of doses to an inadvertent intruder.

The purpose of the screening-level assessment was to provide a conservative estimate of potential drinking water doses to an offsite individual. If for any radionuclide the calculated dose is significantly less than the acceptable dose constraint for this conservative analysis, that radionuclide was not considered further in the assessment. Based on this screening-level assessment, Cook (1998) identified that cesium-135 would need to be considered in a performance assessment of the direct grout option, whereas it had been screened from consideration in the saltstone performance assessment (Martin Marietta Energy Systems, Inc., et al., 1992; Westinghouse Savannah River Company, 1998a). The calculated dose for cesium-135 in this screening assessment was a factor of 30,000 higher than in the screening analysis using the reference salt supernate waste stream concentrations (see Table 7.1 in the main text).

---

[1]An assessment intended to remove radionuclides from further consideration, thus reducing the work load of the analyst.

A simplified groundwater transport analysis was undertaken using the PATHRAE computer code, a rather old code for evaluation of transport in groundwater and its environmental consequences. Several versions of the PATHRAE code exist (e.g., Rogers et al., 1985; Merrell, Rogers, and Bollenbacher, 1986; Rogers and Hung, 1987). Cook (1998) did not specify which version was used, nor how the simple model in PATHRAE was justified as an appropriate screening-level tool for the saltstone disposal facility. This assessment was conducted for the case of a degraded vault (Cook, 1998). Results of the assessment for three key radionuclides are shown in Table E.1, compared to results from the saltstone performance assessment model (Martin Marietta Energy Systems, Inc., et al., 1992; Westinghouse Savannah River Company, 1998a). The peak dose from cesium-135 corresponds to a dose from drinking water of 1.7 mrem yr^{-1}. This value is quite close to the drinking water standard of 4 mrem yr^{-1} in DOE Order 435.1, but Cook (1998) argued that a detailed performance assessment would lead to lower doses from Cs-135 than were found in the PATHRAE analysis.

TABLE E.1 Comparison of Results from the Saltstone Performance Assessment Model and the Simplified PATHRAE Analysis of Cook (1998) for the Direct Grout Waste Stream

Radionuclide	Saltstone Performance Assessment Time of Peak (year)	Saltstone Performance Assessment Peak Concentration (pCi/L)	PATHRAE Time of Peak (year)	PATHRAE Peak Concentration (pCi/L)
Selenium-79	150,000	4.4	20,000	73
Iodine-129	3,200	0.075	21,000	5.9
Cesium-135	negligible	negligible	100,000	330

A full analysis of the intruder scenario was not conducted by Cook (1998). Instead, results were derived from existing results for the saltstone disposal facility, with revised radionuclide inventory values. Intrusion assessment results were not available for cesium-135 from the saltstone performance assessment (Martin Marietta Energy Systems, Inc., et al., 1992) because it had been screened out of the performance assessment for groundwater impacts. For the direct grout assessment, rather than recalculating intrusion scenario results for this radionuclide, results from the E-area vault performance assessment (Westinghouse Savannah River Company, 1998b) were used, and the results were corrected for differing vault volume and inventory.

REFERENCES CITED

Cook, J.R. 1998. Effect of "Grout-it-all" on Saltstone Performance Assessment. Westinghouse Savannah River Company SRT-WED-98-0119, Rev. 2, Aiken, SC.

Martin Marietta Energy Systems, Inc., EG&G Idaho, Inc., Westinghouse Company, and Westinghouse Savannah River Company. 1992 (December 18). Radiological Performance Assessment for the Z-Area saltstone disposal facility. Westinghouse Savannah River Company WSRC-RP-92-1360, Aiken, SC.

Merrell, G.B., V.C. Rogers, and M.K. Bollenbacher. 1986. The PATHRAE-RAD Performance Assessment Code for the Land Disposal of Radioactive Wastes. Rogers and Associates Engineering Corporation RAE-8511-28, Salt Lake City, UT.

Rogers, V.C., G.M. Sandquist, G.M. Merrell, and A. Sutherland. 1985. The PATHRAE-T Code for Analyzing Risks from radioactive Waste. Rogers and Associates Engineering Corporation RAE-8339/12-2, Salt Lake City, UT.

Rogers, V.C., and C. Hung. 1987. PATHRAE-EPA: A Low-Level Radioactive Waste Environmental Transport and Risk Assessment Code. U.S. Environmental Protection Agency EPA 520/1-87-028, Washington, D.C.

Westinghouse Savannah River Company. 1998a. Addendum to the Radiological Performance Assessment for the Z-Area Saltstone Disposal Facility. WSRC-RP-98-00156, Rev. 0, Aiken, SC.

Westinghouse Savannah River Company. 1998b. Radiological Performance Assessment for the E-Area Vaults Disposal Facility. WSRC-RP-94-218, Rev 1, Aiken, SC.

Appendix F

Defense Nuclear Facilities Safety Board
Recommendation 96-1 to the Secretary of Energy

pursuant to 42 U.S.C.§ 2286a(a)(5)
Atomic Energy Act of 1954, as amended

Dated: August 14, 1996

The Defense Nuclear Facilities Safety Board (Board) has devoted substantial attention to the planned use of the In-Tank Precipitation (ITP) System at the Savannah River Site, because of its importance to removal of high-level radioactive waste from storage tanks at that Site, and because certain unique hazards are associated with the ITP process.

The hazards are a consequence of the volatile and flammable organic compound benzene that is released during the process in amounts that must not exceed safe limits. The benzene is generated through decomposition of tetraphenylborate (TPB) compounds. These compounds are added in the process with the objective to precipitate and remove radioactive cesium from solution in the waste water destined for the saltstone process. The concentrated slurry containing the precipitated cesium constitutes a much smaller volume than the original waste, and its feed to the vitrification process leads to production of a correspondingly smaller amount of glass ultimately to be disposed of in a repository.

The proposed treatment process calls for addition of a quantity of TPB in excess of that theoretically required to precipitate the cesium as cesium TPB. That excess is required partly because the significant amount of potassium present is also precipitated as potassium TPB, and partly because an excess of TPB in solution ensures more effective scrubbing of the radioactive cesium through precipitation. However, the benefit of effective scrubbing is accompanied by the generation of the benzene, which presents hazards of a different sort, and which also requires safety controls.

Westinghouse Savannah River Company is the Department of Energy contractor in charge of ITP. The Westinghouse staff at the Savannah River Site believed until recently that the principal cause of decomposition of TPB and generation of benzene is exposure of the TPB to the high level of radiation in the waste. That belief was based on results of full-scale tests conducted in 1983 that may have been misinterpreted, and on a decade of subsequent bench-scale tests using non-radioactive simulants (almost exclusively) rather than actual waste. The first large-scale operations with actual waste since 1983 were conducted recently in Tank 48, and they showed that the generation and release of benzene did not follow predictions. The generation of benzene in the waste under treatment in Tank 48 was unexpectedly rapid. A surprisingly large amount of the benzene remained captured in the waste, and that benzene was released through action of mixing pumps in the tank.

The current view of the contractor staff is that benzene is produced principally through catalytic decomposition of TPB ions in solution. They believe the catalysts are potentially both soluble and insoluble species, one of which is soluble copper known to be present in the waste. They also believe that the cesium TPB precipitate and the potassium TPB precipitate are relatively immune to catalytic decomposition. The contractor proposes to conduct two Process Verification Tests (PVT), PVT-1 and PVT-2, to further establish the validity of these views and to demonstrate the accuracy of the model it has developed to predict the rate at which the captured benzene is released from solution. PVT-1 would be performed on the homogenized nuclear waste now in Tank 48, which has already been treated with TPB that subsequently has partly decomposed with the result that some cesium has returned to solution. Additional TPB would be added to this material to reprecipitate the cesium. The amount of TPB to be added would be strictly limited to a small amount as needed to reduce the concentration of cesium remaining in solution to a low radiation level acceptable for processing as low level waste in the saltstone process, and a large part of that solution would be sent to saltstone. The subsequent proposed experiment, PVT-2, will involve adding to the slurry remaining in Tank 48 a large amount of additional untreated waste and a substantial quantity of TPB as needed to precipitate the cesium in this new waste.

The Board has been informed that the primary safety precaution for the proposed cesium removal activities is to maintain an inert atmosphere in the headspace of Tank 48. This is to be done through establishing a sufficient flow of nitrogen to the tank. Two nitrogen feed systems are available, a normal system and a supplemental emergency system. The nitrogen systems are present to keep the concentration of oxygen below the level that would support combustion of the benzene. Westinghouse staff members have pointed out that these redundant inerting systems provide a sufficient safety factor for control of oxygen concentration in the headspace. They have further stated that the rate of buildup of oxygen concentration from air ingress into the tank headspace, if both inerting systems are simultaneously inoperable, would be slow enough to allow reestablishment of nitrogen flow before the bulk vapor in Tank 48 reaches the minimum oxygen concentration that could support combustion of benzene.

Operations since December 1995 indicate that for the current batch of waste, mixing pump operation increases the benzene release rate from the waste and that turning off the pumps essentially stops the release. The Board has been informed of the consequent belief that the actual rate of benzene release into the tank's headspace and its subsequent removal can be controlled through managing the action of the mixing pumps. This stratagem is to be followed in the tests as a means of maintaining the concentration of benzene in the headspace at a low enough level to prevent it from becoming flammable even if the oxygen concentration were to increase to an undesired level.

Westinghouse representatives also plan to impose a temperature limit for PVT-1 which is expected to prevent decomposition of TPB or to reduce its rate. Finally, they state that for PVT-1 the addition of TPB will be limited to 200 gallons of fresh 0.5 Molar sodium TPB solution, and that any subsequent additions during this experiment would be subject to review and approval by the Department of Energy. Westinghouse believes that this, in turn, would limit the maximum amount of additional benzene that could be produced. In effect, the amount of TPB added will be treated as an Operating Limit.

The Department and its contractor have brought substantial expertise to bear on understanding the science of the ITP process and the phenomena attending it. However, the Board is concerned that some important questions remain unanswered. First, the physical basis for holdup of large amounts of benzene in the waste and its removal through mixing pump operation is not yet well understood. Therefore, confidence in the ability to control its release is not as high as desired.

The Board is also concerned with the results of a recent laboratory-scale experiment using Tank 48 solution and TPB additive. The results from this experiment indicate that the amount of TPB which decomposed exceeded that amount which had been added during the experiment, suggesting that the cesium and potassium TPB precipitates had also partially decomposed, presumably through catalytic attack. If the cesium and potassium TPB precipitates were subject to rapid and extensive attack by a catalyst, an enormous amount of benzene could be generated, and the rate of release could be rapid enough to overwhelm the removal capability of the purging system for Tank 48.

The Board concurs with the view that ITP is of high value for subsequent vitrification of the nuclear waste in the tanks at the Savannah River Site, and that further testing is necessary to gain a better understanding of the science of the process to assure safety during and after precipitation of the cesium.

The Board believes that if it were conducted according to the limitations stated above, PVT-1 can be run safely and can help in leading to an improved understanding of the science and the mechanisms involved in the ITP process.

The present plan for conduct of PVT-2 involves new and untested nuclear waste and a much larger addition of TPB. Furthermore, the liquid in Tank 49, which contains TPB from the previously mentioned 1983 demonstration test, is to be used as the source of a significant part of the TPB to be added to Tank 48 during PVT-2. The Board understands that Tank 49 was also the source of TPB used in the one experiment which led to an apparent decomposition of precipitated cesium and potassium TPB. One very probable interpretation of that anomaly is that the material in Tank 49 contains an unknown catalyst which can attack the precipitated material and might also increase the rate of release of benzene by an amount that is unpredictable at present. Furthermore, waste from tanks not yet tested could contain unknown constituents that could also adversely affect the rate of production and release of benzene.

The Board believes that the uncertainty in understanding of the science of ITP would make it imprudent to proceed from PVT-I to PVT-2 without substantial improvement in the level of understanding. Some such improvement may follow interpretation of the results of PVT-I. Better understanding of the anomalous experiment suggesting decomposition of TPB precipitates is also required.

Therefore, the Board makes the following recommendations:

- Conduct of the planned test PVT-2 should not proceed without improved understanding of the mechanisms of formation of the benzene that it will generate, and the amount and rate of release that may be encountered for that benzene.
- The additional investigative effort should include further work to (a) uncover the reason for the apparent decomposition of precipitated TPB in the anomalous experiment, (b) identify the important catalysts that will be encountered in the course of ITP, and develop quantitative understanding of the action of these catalysts, (c) establish, convincingly, the chemical and physical mechanisms that determine how and to what extent benzene is retained in the waste slurry, why it is released during mixing pump operation, and any additional mechanisms that might lead to rapid release of benzene, and (d) affirm the adequacy of existing safety measures or devise such additions as may be needed.

John T. Conway
Chairman

Appendix G

Acronyms and Abbreviations

ANL	Argonne National Laboratory
CST	crystalline silicotitanate
CsTPB	cesium tetraphenylborate
DF	decontamination factor
DNFSB	Defense Nuclear Facilities Safety Board
DOE	United States Department of Energy
DWPF	Defense Waste Processing Facility
EM	DOE Office of Environmental Management
HLW	high-level waste
ITP	in-tank precipitation
IPET	Independent Project Evaluation Team
KTPB	potassium tetraphenylborate
MST	monosodium titanate
NaTPB	sodium tetraphenylborate
NRC	National Research Council
OST	Office of Science and Technology
ORNL	Oak Ridge National Laboratory
PCMP	Process Chemistry and Mechanisms Panel
PHP	Precipitate Hydrolysis Process
RD&D	research, development, and demonstration
R&D	research and development
SCDHEC	South Carolina Department of Health and Environmental Control
SNT	sodium nonatitanate
SRS	Savannah River Site
SRTC	Savannah River Technology Center
TPB	tetraphenylborate ion $[B(C_6H_5)_4]^-$
TRU	transuranic
USNRC	United States Nuclear Regulatory Commission
WSRC	Westinghouse Savannah River Company